THE OSHA ANSWER BOOK

D1536434

Published By:

MORAN ASSOCIATES
1600 Brighton Bluff Ct.
Orange Park, FL 32073
(904) 278-5155

ISBN: 0-9632296-7-2

Library of Congress Cataloging in Publication Data

Moran, Robert D.   1929 - 1994
Moran, Mark McGuire 1959 -

THE OSHA ANSWER BOOK
Third Edition, 1996

1. Industrial safety - Law and legislation - United States.
2. Industrial hygiene - Law and legislation - United States.
3. United States. Occupational Safety and Health Administration.
I. Title

TABLE OF CONTENTS

THE OSHA ANSWER BOOK

INTRODUCTION

This book is an effort to place everything that most employers need to know about OSHA into a single volume. It is a reference book, designed to be consulted frequently. Information on OSHA that every employer should have is contained within the pages of the book itself. It also provides the name, address and telephone number of government-funded OSHA experts and consultants in every state who will provide additional information employers may need.

This book is divided into four parts:

PART ONE explains what OSHA is, how it's enforced, how employers can cope with OSHA enforcement, and the consequences of non-compliance.

PART TWO explains in some detail the numerous OSHA standards and regulations with which employers must comply.

PART THREE is a state-by-state OSHA directory that lists each local OSHA enforcement office and each government-funded consultative service.

The OSHA consultants listed in Part Three exist solely to help employers achieve compliance with OSHA requirements. They are paid by the government. There is no charge for the services they provide to private employers. They will answer OSHA questions when asked and they make house calls.

The consultants are experts in OSHA compliance. If you want them to, they will come to your premises, survey your worksite, tell you whether or not you're in compliance with OSHA and, if not, will explain what you must do to obtain compliance. A more detailed explanation of the consultative service begins in this section.

PART FOUR is safety on the Internet. It will help anyone find information on the Internet. It lists all the U. S. Government web sites (including OSHA), state government web sites, and safety and health web sites.

HOW TO USE THIS BOOK

Every employer who is obligated to comply with the Occupational Safety and Health Act should read Part One. That should provide the reader with an incentive to read at least the first section of Part Two and to run-through those OSHA requirements listed in the remaining sections of Part Two that apply in his/her own business. That exercise may raise questions or doubts about the employer's OSHA compliance status. If so, Part Three tells you who to contact for the answer.

If you want to find information on any subject or area of safety, you should refer to Part Four. There is so much information on the Internet you may never come back!

APPLICABILITY OF OSHA STANDARDS

Not all OSHA standards apply to all employers. Those that are codified in Title 29 of the Code of Regulations (C.F.R.) at Parts 1903 and 1904 (injury/illness recordkeeping and general reporting requirements) apply to most employers. The OSHA standards in the remaining parts of Title 29 are ostensibly limited to the areas of business and industry suggested by their title: Part 1910 - general industry, Part 1915 - shipyards, Part 1926 - construction, Part 1928 - agriculture. However, that is not always the case.

Some of the Part 1910 "General Industry" standards also apply in the construction, agriculture, and maritime industries. The rule of thumb applied by OSHA is this: *If an OSHA inspector observes a hazardous condition at a construction (or maritime) worksite but there is no construction (or maritime) standard regulating that condition, he will cite you under the "general industry" standard that regulates that condition.* The authority for that OSHA practice is stated in 29 C.F.R. §1910.5(c): *any standard shall apply according to its terms to* any *employment and place of employment in any industry, even though particular standards are also prescribed for that industry.*

Thus, OSHA can cite an employer in the maritime industry or the construction industry for noncompliance with a "general industry" (Part 1910) standard in those situations where there is **no** applicable maritime or construction standard that regulates the condition in question.

In most cases, the applicability of an OSHA standard can easily be ascertained from its text. In some cases, however, additional research may be necessary. For example, the particular "general industry" standard requiring records for fixed fire extinguishing systems is 29 C.F.R. §1910.160(b)(9). That is a 2-sentence standard that, if taken literally, applies to all fire extinguishing systems. However, §1910.160(b)(9) is a subsection of §1910.160. That standard begins with a section entitled *"scope and application,"* §1910.160(a)(1), which states, for example, that the standard does **not** apply to automatic sprinkler systems.

Many other OSHA standards are structured similarly. They begin with *"scope and application"* provisions. Regrettably, however, not all OSHA standards clearly specify their applicability. Further research -- or a telephone call to the nearest consultative service -- may be needed to determine the applicability of those standards.

THE OSHA STANDARDS: HOW TO GET COPIES

There should be a copy of the OSHA standards at every workplace. Indeed, a number of standards include requirements that the text of the standard **must** be present at the workplace so that it can be consulted by employees at their convenience. You will be in violation of those standards if you do not have a copy **at each place where employees work**.

The standards are bound in paperback books (but are also available on computer disks). If you want all OSHA standards and regulations in book form, you will have to get six separate volumes, but if you only want the construction standards or the maritime standards, you'll only need one book (general industry standards cover 2 books).

The OSHA standards can be purchased from the U.S. Government Printing Office directly or from one of its bookstores. Their phone number is (202) 512-1801. There are also some private publishers who sell copies of the OSHA standards in book (and computer disk) format. Contact Government Institutes at (301) 921-2355. In addition, some of the state and local OSHA offices will provide a copy upon request. Part Three of this book contains a list of the local OSHA offices.

THE PURPOSE OF THIS BOOK

The purpose of this book is to supply employers with answers to OSHA problems, and to help solve many of the unknown OSHA requirements.

The price of the book has been kept as low as feasibly possible so that all employers can afford to keep a copy at every place where employees work.

We try to keep this book updated whenever new regulations change or information needs to be revised. We hope this edition is helpful and useful for you.

PART ONE

THE OSHA ENFORCEMENT SCHEME

I. INTRODUCTION TO OSHA

1. Who Is Obligated To Observe The Law

If you have one or more employees, you are an employer. If you are an employer, you are obligated to comply with the Occupational Safety and Health Act (OSH Act). There are very few exceptions. Coal mines are excepted because they are regulated by their own OSHA-type law. The same goes for government agencies. Everyone else must comply.

It doesn't just apply to factories and construction sites -- or only to hazardous jobs. It applies to all employers. Office work is covered. Retail stores are covered. If you employ anyone to do any kind of work, you must observe the OSH Act.

OSHA requirements apply even if there are no accidents or injuries. The purpose of the OSH Act is to prevent the **first** injury. Thousands of employers with perfect safety records have been penalized by OSHA. Other employers who have **never** been inspected by OSHA have been found liable in civil liability court cases because they had failed to observe an OSHA requirement.

You must observe the OSHA requirements even though you are unaware of any hazardous conditions. It can be compared with the income tax rule that requires you to file a return even if you don't owe any taxes.

Employers are cited even when the OSHA violation is an employee's fault, such as failure to wear a hard hat, safety shoes or goggles. OSHA places the onus on employers.

The law provides that each employer is supposed to make sure that his employees observe all safety requirements.

There is no small business exception. Employers with 10 or fewer employees don't have to keep injury/illness records and are exempted from some OSHA inspections, but they must comply with all OSHA safety and health standards and requirements.

2. What The Law Requires

There are thousands of OSHA requirements. They include:

* Fire Protection
* Electricity Rules
* Sanitation and Air Quality Requirements
* Machine Use, Maintenance & Repair
* Posting Notices and Warnings
* Reporting Accidents and Illnesses
* Keeping Detailed Records
* Adopting Written Compliance Program
* Employee Training and Qualifications

That's just a sample. There are several thousand others.

OSHA violations occur when you do not observe all the requirements that apply. Simply running a safe operation is not enough. And ignorance of the law is no excuse. The penalties for noncompliance are severe. Million dollar fines are not unusual. But that's not all! You could go to jail or be sued for millions even if you have never received an OSHA inspection.

Here's an example:

William Boss hired Joe Workman to do some painting. Boss didn't know that Workman had a slight asthma condition. Working with the paint severely aggravated Workman's condition. He had to be hospitalized and nearly died. He can never work again. Boss was sued for ten million dollars because he failed to heed the material safety data sheet (MSDS) warnings for the paint. They provided that no one should use the paint unless first given a pulmonary function test. The fact that Boss didn't know about the MSDS warnings or Workman's asthma condition was no defense. There is an OSHA requirement that an MSDS must be obtained for all products that contain hazardous chemicals and that each employer must provide training to his employees on the MSDS provisions. Boss hadn't done that.

Everyone can learn a lesson from this example. If you are required to provide protection, give a warning, provide employee training, or use particular compliance methods, and you **HAVE NOT COMPLIED** because **YOU DIDN'T KNOW ABOUT IT**, the consequences could be severe. If someone is hurt or becomes disabled as a result, you could face criminal charges or a multi-million dollar liability suit, or both. And your defenses may be weak because ignorance of the law is no excuse. Of course, there could also be sanctions imposed by OSHA -- even if no one was hurt.

The purpose of this book is to help you avoid problems of that kind. It will provide you with a general idea of what OSHA requires and tell you where to write or telephone for more detailed answers and free on-site consultation.

II.   ENFORCEMENT

1.   How OSHA Is Enforced

There are approximately 2000 OSHA inspectors stationed in various places across the country. They conduct unannounced inspections of places where people are at work. If they discover conditions or practices that they think are in violation of OSHA requirements -- and they nearly always do -- citations and penalties against the employer will soon follow.

About half the inspectors are federal employees. The others are state employees. The people that run the OSHA program where you work depends on the state you are in. The federal government enforces OSHA in 29 of the 50 states. State government employees run the program in the other 21 -- and in 2 U.S. Territories (Puerto Rico and the Virgin Islands). Those 23 states and territories are known as "State Plan" states. They are listed by name on the following map:

State Plan States

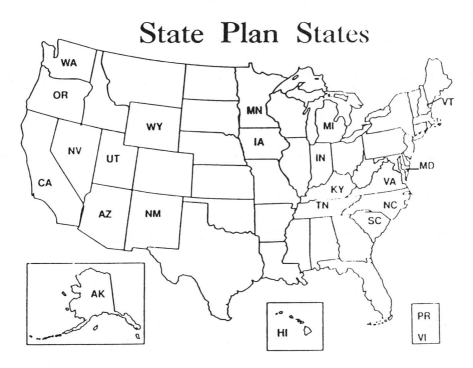

The federal government inspectors are part of OSHA -- the U.S. Labor Department's Occupational Safety and Health Administration. The state inspectors have the same power as OSHA inspectors and, for the most part, they enforce the same requirements.

OSHA has delegated its enforcement authority to the state government in the so-called "State Plan" states. There are 23 of them. A list follows in alphabetical order:

Alaska

Arizona

California

Hawaii

Indiana

Iowa

Kentucky

Maryland

Michigan

Minnesota

Nevada

New Mexico

North Carolina

Oregon

Puerto Rico

South Carolina

Tennessee

Utah

Vermont

Virgin Islands

Virginia

Washington

Wyoming

2.   How Inspection Targets Are Chosen

The particular places that OSHA people inspect are chosen according to an OSHA "inspection plan (or program)." There are several different ones. They fall into one of three general categories that will be explained below:

- Programmed

- Fatality/catastrophe

- Complaint/referral.

A.   Programmed Inspections

Approximately 3 out of every 4 OSHA inspections are "programmed." That means that the place to be inspected (the "establishment") is chosen at random from a list (or lists) that OSHA has made up. Your company name is probably on that list.

The list of all business establishments is divided into industrial categories according to standard industrial classification, or "SIC code" numbers.

> It should be noted that, in OSHA lexicon, the word "establishment" is a word of art. Those who don't realize that fact often confuse it with terms such as business or industry. It is important, therefore, to recognize and understand that difference. For OSHA purposes the word "establishment" means: a single physical location where business is conducted or where services or industrial operations are performed. For example, all hotels in the country would be classified as the "hotel industry" (SIC 7011). An individual hotel within that industry, such as the Palmer House or the Logan Airport Holiday Inn, would be an "establishment" within the hotel industry.

Whether you know it or not, your business is part of an "industry" which has an "SIC code" number. Each industry has a code number depending on the type of work it does. For example:

- Bakeries: SIC Code 2051
- TV Repair Shops: SIC Code 7622
- Nonferrous Foundries: SIC Code 3362

There are over 10,000 such categories. SIC code numbers existed long before OSHA existed. They were developed for statistical and research purposes and are widely used by both government and business.

The U.S. Department of Labor keeps track of employee injury and illness statistics based on SIC codes. An annual survey is made by the Bureau of Labor Statistics. That survey will list each industry by SIC code and report the employee injury/illness rate for each of them.

OSHA then uses those figures to program their inspections. Each code number is assigned a priority based on the past injury/illness experience of that particular industry. OSHA tries to follow those priorities and select businesses for inspection on a "worst first" basis (but it doesn't always succeed).

Here's an example.

> If 1992 statistics showed that the retail bakery industry had the worst record for employee injuries, all retail bakeries (SIC Code 2051) would be Number One on OSHA's 1993 list of inspection priorities. If there were 25 bakeries in Cleveland, all 25 would be inspected by Cleveland OSHA people at the beginning of the year. All other OSHA offices in the country would do the same for their own local bakeries. After the local OSHA office had completed their inspection of all bakeries, they would then go to the industry listed as Number Two on their priority list and inspect all local establishments in that industry. That process would continue in sequence.

OSHA will sometimes select certain industries (by SIC code) for "special emphasis" and give them a higher priority. In the past that has happened for trenching and excavation contractors, fireworks manufacturers, foundries and a number of others.

There is no way of telling where you stand on the list or when (or if) you will be chosen for a "programmed" inspection. Indeed, it is a criminal offense for anyone to provide such information.

Section 17(f) of the OSH Act provides that: *"Any person who gives advance notice of any inspection to be conducted under this Act, without authority from the Secretary [of Labor] or his designees, shall, upon conviction, be punished by a fine of not more than $1,000 or by <u>imprisonment</u> for not more than six months, or by both."*

B. **Fatality/Catastrophe Inspections**

OSHA will nearly always inspect a place where an employee has been killed on the job, where there have been multiple employee hospitalizations, or there has been a newspaper, radio or TV report of an accident, explosion, spill or calamity of some kind. An OSHA inspector will probably show up shortly after he learns of it.

C. **Complaint/Referral Inspections**

Employees (particularly recently-discharged employees) and labor unions can (and often do) complain to OSHA about conditions at particular worksites. Unions will file complaints even if they don't represent the employees. That sometimes happens when they are trying to organize the employees or because they are hostile to a particular company.

Those complaints frequently trigger a special OSHA inspection.

Inspections are also triggered by "referrals". Those could come from OSHA inspectors themselves; doctors who have treated employees; local, state or federal employees; disgruntled salesmen; competitors; relatives of present or former employees; environmental, neighborhood, and civic activists; or almost anyone. OSHA doesn't follow up with an inspection on **all** such "tips;" but some of them do instigate inspections.

If the inspection is triggered by an employee complaint, the OSHA inspector is required to provide you with a copy of the complaint (minus the employee's name) no later than at the time of inspection. But in many cases, there is no way you can tell how you came to be targeted by OSHA.

D.   Focused Inspections in General Industry

Under previous agency policy all inspections were comprehensive in scope, addressing all areas of the workplace. This caused compliance officers to spend too much time and effort on a few employers looking for all violations and, thus, too little time overall on inspections for hazards which are most likely to cause fatalities and serious injuries to workers.

The goal of OSHA's focused inspections is to have the Compliance Safety Health Officer's (CSHO's) time more effectively spent inspecting the most serious hazardous workplace conditions. The OSHA inspector will conduct comprehensive inspections only on those employers where there is inadequate commitment to safety and health.

This will be what OSHA calls a "pilot study." This means it is being tested. It is **not** a standard or even a nationwide policy at this point. It will be tested in six field offices: (1) Harrisburg, PA; (2) Austin, TX; (3) Billings, MT; (4) Cincinatti, OH; (5) Atlanta, GA (east); and (6) Atlanta, GA (west). It will be using the Program Evaluation

Profile (PEP). The safety ratings derived from the evaluation will be factored into a new formula for adjusting penalties, also to be given a tryout.

During PEP's trial run, OSHA inspectors will rate each of these program components on a scale of 1 (absent or ineffective) to 5 (outstanding), helped by 15 tables listing descriptors of workplace characteristics that best fit what they see in the field. Using these categories and ratings tables, you can do the same thing in your own facility.

This is how it works: The policy will have six core elements that will be scored. Management leadership and employee participation, workplace analysis, accident and record analysis, hazard prevention and control, emergency response, and training. These six main elements will then be further divided into 15 components, or what OSHA calls factors. Each factor receives its own rating, and the score for each of the six elements is determined by the average of the element's factor scores.

The score for the management leadership and employee participation element (which OSHA considers the foundation of a program) will be based on whichever is the lowest of the following: the score for management leadership; the score for employee participation; or the average score for all four factors. The sixth element, training, has no separate factors and so receives just one score.

The **overall score** for a work site will be the average of the six scores for core elements, rounded to the nearest whole number.

If an element or factor does not apply to a work site, it will be noted as "not applicable" and will not affect the score.

During all inspections CSHO's will determine whether or not there is a safety coordinator and then proceed to determine the following:

- Whether or not an adequate safety and health program meets the guidelines set forth, and
- A designated competent person responsible for implementing the program.

If the employer meets both of the criteria above, then a focused inspection will be made. If the employer does not have a safety and health program, then the inspection will proceed in accordance with previously established procedures for a comprehensive inspection.

The OSHA Inspector will assess the company's safety and health program and will consider the following:

- The comprehensiveness of the program;
- The degree of program implementation;
- The designation of competent persons;
- How the program is enforced, including management policies, employee involvement, training and;
- Employees will also be interviewed during the walkaround in the evaluation of the safety and health program.

Employers will be chosen for the focused inspection policy based on such factors as their high accident, illness and fatality rates.

When an industry is targeted for focused inspections, early efforts will be made to let them know that OSHA is conducting a program that will affect them.

OSHA's Regional Administrators will use suitable means to advise employers in their Regions within SICs designated for focused inspections that enforcement activity is planned and to offer information to help them comply.

OSHA's intention is that employers in affected SICs be notified early enough so that they have an opportunity to implement appropriate safety and health programs before enforcement activities begin. Such employers should be given the opportunity to become informed of the criteria against which their programs will be evaluated.

Focused inspections will be conducted in programmed safety inspections in designated SIC codes in general industry. Such inspections will have the following procedures:

1. In the course of obtaining information to complete the Program Evaluation Plan (PEP), Form OSHA-195 the compliance officer will explain to the employer that:

    - Each inspection begins at the opening conference without an assumption as to whether the employer will or will not quality for a focused inspection;
    - The inspection may focus on the defined group of serious hazards, if an effective safety and health program is found to be implemented; and
    - The compliance officer's initial assessment on the PEP will be verified - or modified - based on information obtained in interviews of an appropriately representative number of employees and by observation of actual safety and health conditions during the inspection process.

2.   Each inspection begins as a comprehensive inspection, until a determination is made, in accordance with the following guidance, to "focus" it. The compliance officer may begin the walkaround phase of the inspection with the intention of limiting its scope to the designated focus hazards if the employer has:

- A safety and health program that appears to have an overall score (level) of 2 or higher according to the PEP, and
- Designated person(s) responsible for and capable of implementing the program.

3.   The compliance officer may continue to limit the inspection to the designated focus hazards for the industry as long as the evidence observed during the conduct of the inspection continues to support a finding that the employer's safety and health program is being effectively implemented. If, however, the number of hazards found in the workplace indicates that the employer's program is inadequate, no *focused* inspection will be conducted.

4.   Citations will be issued for all serious violations that are observed (whether or not they relate to the focus hazards) and for all other-than-serious violations that are not abated by the employer before the end of the inspection. If the inspection is focused, other-than-serious violations that are abated immediately in the presence of the compliance officer will not be cited.

The compliance officer's evaluation of the safety and health program contained in the PEP will be shared with the employer and with employee representatives, if any, no later than the date of issuance of citations (if any).

The preliminary assessment from the PEP will be discussed with the employer in the closing conference; the employer will be advised that this assessment may be modified based on further inspection results or additional information supplied by the employer.

The PEP will be used as a source of safety and health program evaluation for the employer, employees, and OSHA. It will be used during the opening conference and throughout the inspection process. It will also be used for the following:

- The compliance officer will explain the purpose of the PEP and obtain information about the employer's safety and health program in order to make an initial assessment about the program.

- The initial assessment will be verified - or modified - based on information obtained in interviews of an appropriately representative number of employees and by observation of actual safety and health conditions during the inspection process.

- If the employer does not wish to volunteer the information needed for the PEP, the compliance officer will note this in the case file but will not press the issue. The benefits of a PEP evaluation will, however, be explained.

17

These elements (except for number 6 *Training*) are divided into factors, which will also be scored. The score for an element will be determined by the factor scores. The factors are:

1. *Management Leadership and Employee Participation.*

    - Management leadership
    - Employee participation
    - Implementation (tools provided by management, including budget, information, personnel, assigned responsibility, adequate expertise and authority, line accountability, and program review prodecures).
    - Contractor safety

2. *Workplace Analysis*

    - Survey and hazard analysis
    - Inspection
    - Reporting

3. *Accident and Record Analysis*

    - Investigation of accidents and near-miss incidents
    - Data analysis

4. *Hazard Prevention and Control*

    - Hazard control
    - Maintenance
    - Medical program

5. *Emergency Response*

    - Emergency preparedness
    - First aid

6.   *Safety and Health Training*

- Classroom training
- Video tapes, books, etc.

The compliance officer will score the employer on each of the individual factors and elements after obtaining the necessary information to do so. These will be given a score of 1, 2, 3, 4, or 5.

OSHA inspectors will determine the scores for each of the six elements as follows:

- The score for the "Management Leadership" factor.
- The score for the "Employee Participation" factor.
- The average score for all four factors.

The two factors of "Management Leadership" and "Employee Participation" will be given greater weight because they are considered the foundation of a safety and health program.

The sixth element, *Training*, will be determined from the level 1-5 that best fits the worksite and will be noted in the appropriate box on the PEP. For each of the other four elements, **average** the scores for the factors. In **averaging** factor scores, round to the nearest whole number (1,2,3,4, or 5). Round up from one-half (.5) or greater; round down from less than one-half (.5).

If the employer declines to provide pertinent information regarding one or more factors or elements, a score of 1 will be recorded for the factor or element. If the element does not apply to the worksite being inspected, a notation of **"Not Applicable"** will be made in the space provided. This will be represented by **"N/A"** or **"0."** This will not affect the score.

An **"Overall Score"** for the worksite will be recorded on the score summary. This will be the average of the six individual scores for elements, rounded to the nearest whole number (1,2,3,4, or 5). Round up from one-half (.5) or greater; round down from less than one-half (.5).

The six individual element scores, in sequence (e.g., "2-2-1-2-3-1") will constitute a "rating" for purposes of tracking in an employer's safety and health program, and will be recorded.

The Overall Score on the PEP constitutes the "level" at which the company's safety and health program is scored. **Remember: this level is a relatively informal assessment of the program, and it does not represent a compliance judgment by OSHA** - that is, it does not determine whether an employer is in compliance with OSHA standards. Each of the elements and factors will be scored from 1 to 5 indicating the level of the safety and health program. The following chart summarizes the levels:

Score	Level of Safety and Health Program
5	Outstanding program
4	Superior program
3	Basic program
2	Developmental program
1	No program or ineffective program

The following will be taken into account in assessing specific factors:

Employer safety and health programs should be in writing in order to be effectively implemented and communicated as follows:

- A program's effectiveness is more important than whether it is in writing. A small worksite may well have an effective program that is not written, but which is well understood and followed by employees.

- In assessing the effectiveness of a safety and health program that is not in written form, compliance officers should follow the general principles laid out in OSHA "Citation Policy for Paperwork and Written Program Requirement Violations." That is:

  - An employer's failure to comply with a paperwork requirement is normally penalized only when there is a serious hazard related to this requirement.
  - An employer's failure to comply with a written program requirment is normally not penalized if the employer is actually taking the actions that are the subject of the requirement.

- In using the PEP, the compliance officer is responsible for evaluating the employer's actual management of safety and health in the workplace, not just the employer's documentation of a safety and health program.

The importance of a safety and health program's comprehensiveness is implicitly addressed in Workplace Analysis under both "Survey and hazard analysis" and "Data analysis." An effective safety and health program will address all known and potential sources of workplace injuries and illnesses, whether or not they are covered by a specific OSHA standard. For example, lifting hazards and workplace violence problems should be addressed if they pertain to the specific conditions in the employer's workplace.

The PEP evaluation and the scores assigned to the individual elements and factors should be consistent with the types and numbers of violations or hazards found during the inspection and with any citations issued in the case. As a general rule, high scores will be inconsistent with numerous violations or a high injury/illness rate. The following are examples:

- If applicable OSHA standards require training, but the employer does not provide it, the PEP score for "Training" should not normally exceed "2."
- If hazard analyses (e.g., for permit-required confined spaces or process safety management) are required but not performed by the employer, the PEP score for "Workplace analysis" should not normally exceed "2."
- If the inspection finds numerous serious violations relative to the size and type of workplace, the PEP score for "Hazard Prevention and Control" should not normally exceed "2."
- The duration of the PEP review will vary depending on the circumstances of the workplace and the inspection. In all cases, however, this review will include:

  - A review of any appropriate employer documentation relating to the safety and health program.
  - A walkaround inspection of pertinent areas of the workplace.
  - Interviews with an appropriate number of employer and employee representatives.

The PEP interview with the employees or the employer will include these questions:

Background
- What is your job here?
- How long have you worked here?

Management Leadership and Employee Involvement
- Is there an overall policy here regarding employee safety and health? If so, please restate it in your own words or tell me where to find the policy.
- How important is worker safety and health protection to management in this company?
- What have you seen or heard to back up your answer?
- In your own words, what is (are) this year's safety goal(s), and how is it (are they) going to be achieved?
- Does management set a good example when it comes to doing things safely? What (else) does top management do to demonstrate interest in worker safety and health? Is it enough? If not, what else needs to be done?
- Do you have any particular responsibilities for safety and/or health here? If so, what are they? If not, why not?
- What are you expected to do to keep this company safe and healthful? Do you always do it? If not, why not?
- How easy is it to get rid of a safety or health hazard? Give an example.
- What happens when a safey or health goal is not reached?
- If you have had any part in a review of the safety and health program, describe to the best of your knowledge how that review process works.

Worksite Analysis
- Are regular safety and health self-inspections made? Are they done often enough? Do the inspectors know their stuff?
- Have you ever discovered a safety or health hazard? What did you do about it? What should you do about it if you find one tomorrow?
- Are new safety and health precautions introduced at the same time as or after new facilities, equipment, materials, or processes? Give an example.
- What generally happens after an accident? Does investigation turn up root causes? Give an example.

Hazard Prevention and Control
- Do you have a set of instructions that you can go by to keep you working safely? If so, do you follow them?

- If there are any hazardous substances in or around your work area, how are you protected from them? How good is the protection? On what do you base your judgment?
- Is the facility here usually kept reasonably clean?
- Is maintenance done regularly? If you perform any maintenance, do you have a set of instructions to go by?
- Do you know what to do in an emergency? If so, how do you know that, and for what emergencies?
- Is there a medical program here? If so, how does it work? If not, is there a first-aid program here and how does that work?

Safety and Health Training

- Have you been trained by this company on all routine, maintenance, and emergency procedures that you are expected to follow? If so, when, how, by whom, and how good was the training given? If not, did you ask for more training? If so, what reason was given for not providing the training you requested?
- Do you receive regular safety and health training? If so, how much?
- Are you aware of company safety rules? If so, do they seem to cover everything they should?

You can determine the scores for each of the six elements as follows:

- The score for the *Management Leadership and Employee Participation* element will be whichever is the lowest of the following:
    - The score for the "Management Leadership" factor, or
    - The score for the "Employee Participation" factor, or
    - The average score for all four factors.
- For the sixth element, *"Training,"* just determine the level 1-5 that best fits the worksite and note it in the appropriate box on the PEP.
- For each of the other four elements, **average** the scores for the factors.
- If the employer declines to provide pertinent information regarding one or more factors or elements, a score of "1" shall be recorded for the factor or element.
- If the element or factor does not apply to the worksite being inspected, a notation of **"Not Applicable"** will be made in the space provided. This will be represented by **"N/A"** or, a **"0."** This will not affect the score.

An **"Overall Score"** for the worksite will be recorded on the PEP. This will be the **average** of the six individual scores for elements, rounded to the nearest whole number (1,2,3,4, or 5). Round up from one-half (.5) or greater; round down from less than one-half (.5).

> **EXAMPLE**: A PEP's element scores are:
>
> $2.5 + 2.7 + 2.3 + 3.0 + 2.3 + 2.0 = 14.8$
>
> $14.8 \div 6 = 2.467 = 2$ PEP OVERALL SCORE

As the example demonstrated, an employer would have received a "2" as an overall PEP score. This is considered by OSHA as an acceptable safety and health program.

OSHA'S PROGRAM EVALUATION PROFILE FORM

PEP Program Evaluation Profile Employer: Inspection No.: Date: CSHO ID:		Management Leadership and Employee Participation				Workplace Analysis			Accident and Record Analysis		Hazard Prevention and Control			Emergency Response		Safety and Health Training
		Management Leadership	Employee Participation	Implementation	Contractor Safety	Survey and Hazard Analysis	Inspection	Reporting	Accident Investigation	Data Analysis	Hazard Control	Maintenance	Medical Program	Emergency Preparedness	First Aid	Training
Outstanding	5															
Superior	4															
Basic	3															
Developmental	2															
Absent or Ineffective	1															
Score for element																
Overall Score																

OSHA-195 (3/96)

III.   OSHA INSPECTIONS, CITATIONS AND FINES

1.   OSHA's Arrival At Your Business

An OSHA inspector will arrive without warning at your place of business (sometimes it will be two or more inspectors). He will introduce himself, show you his government credentials (I.D. card) and announce that he is there to conduct an OSHA inspection. The manner in which you conduct yourself from that moment on will determine the number and severity of the citations and fines you will receive when the inspection is over. So be careful. You should always observe the following three precautions throughout the inspection:

---

1. An employer representative should stay with the inspector at all times. It should always be the <u>same</u> person. Two or more employer representatives could provide conflicting information. That could mean trouble. The inspector might think your company is trying to mislead him. Management should speak with <u>one</u> <u>voice</u> when dealing with OSHA.

2. Your representative must be very careful what he says. Read the remainder of this section of the book for further explanation of how your statements to OSHA can cause you big trouble. Think before you speak. When unsure about what to say call a "time out" and get help.

3. Be very careful about company records and documents. They may seem innocuous but they frequently can be "smoking guns". Be sure to read the discussion "Cooperation and Corroboration," Part One of this book, Section V. 8., before you discuss any company records or documents with OSHA.

The inspector does **not** have the **right** to inspect. You can refuse to admit him. You don't have to give a reason. You can simply say that your company does not permit warrantless OSHA inspections.

If that happens, the inspector will leave without argument. OSHA must then decide whether to abandon the inspection or apply for an inspection warrant from a nearby magistrate that orders you to permit OSHA to enter and inspect. Usually -- but not always -- OSHA will apply for an inspection warrant when the employer refuses to permit the inspection.

If OSHA gets a warrant, the inspector will return with it at a later time and give you a copy. Usually that will be several days later. But it could be only a few hours later or it could be as much as several weeks later.

The warrant is designed to protect **you** against unauthorized government intrusion into your business. Read it. It will specify what the inspectors are allowed to do at your workplace, where they can go and how long they can stay.

Regardless of whether you admit the inspector because he has a warrant or because you grant him your permission to inspect, the inspection that follows will be the same. It will begin with an "opening conference". Then there will be a "walkaround". It will end with a "closing conference". All three of these activities will be explained.

2.   How To Survive an OSHA Inspection

Surviving an OSHA inspection involves three essential elements: Understanding, Preparation, and Responding. Each is discussed below.

A.   Understanding

The Report of The House Committee that sent the occupational safety and health bill to the floor in 1970 included the minority views of 6 Congressmen who labeled the legislation:

 An "authoritarian, penalty-oriented, 'bull-in-the-china-shop' approach."

Although improvements in the legislation were made before its subsequent enactment, there are no words that could more accurately describe the manner in which that Act has been implemented in recent years by the Occupational Safety and Health Administration (OSHA), the Agency created to enforce it.

Few can argue with the Act's noble purpose of reducing occupational hazards that annually result in the injury and illness of thousands of hardworking employees. But its purpose is not being served by OSHA's cops-and-robbers method of enforcement.

It's doubtful that occupational injuries and illnesses can be significantly reduced by any government regulatory system. Numerous studies have demonstrated that approximately 80% of work-related casualties result from deliberate misconduct, negligence, habits of work and human error, causes that are not -- and cannot be --

controlled by government regulation. For example, the second leading cause of worker fatalities in the U.S. is homicide. Motor vehicle accidents are first. In 1991, both Texas and Massachusetts reported more homicides on the job than motor vehicle accident fatalities.

OSHA inspections do not change a company's worker injury rate. Companies that have been inspected by OSHA and found to have no violations have worker injury/illness rates that are indistinguishable from similar companies that have never experienced an OSHA inspection, as well as others that have been cited for hundreds of OSHA violations. Indeed, OSHA adopted a Voluntary Protection Program (VPP) during the Reagan Administration that exempts certain companies from programmed OSHA inspections. The annual injury/illness rate for the companies that opted for the exemption is substantially lower than comparable companies that are not exempt from OSHA inspection. So much for the notion that OSHA inspections serve a worthwhile purpose.

There are currently more than 2,000 OSHA inspectors -- some Federal, some state. None of them is familiar with all the machines, systems, operations and work processes they are called upon to inspect. But their superiors evaluate their job performance on the number and severity of the citations and penalties that they issue so, not surprisingly, they issue a lot of them. As a result, many such citations call for unnecessary changes in work conditions and practices.

The inspectors know that OSHA has not succeeded in its injury/illness reduction mission. They think the fault lies with employers and with their own Agency's failure to put more employers in jail and penalize them more severely. Those view appear in a November 1990 General Accounting Office report: "Inspectors' Opinions on Improving OSHA Effectiveness."

Although there are some exceptions, OSHA inspectors often exhibit an anti-business, pro-union bias. They have been known to accept misinformation and fabrications from employees without any attempt at verification.

OSHA inspectors get help from the AFL-CIO on their personal wages, hours and working conditions. The Labor Department's management engages in collective bargaining with its employees on those subjects. The OSHA inspectors are represented in that process by an AFL-CIO union.

It's not unusual for draconian OSHA inspections to be scheduled to coincide with a union campaign against a company or for OSHA to cite the company for allegedly violative conditions and practices pointed out to the inspector by a union representative. Frequently those citations have little or nothing to do with employee safety or health.

Employers who understand the foregoing will be better equipped to respond to an OSHA inspection. Many employers, however, conduct themselves during an OSHA inspection as though they believed in the theory that if you're nice to a crocodile, you'll be the last one eaten. Paradoxically, the record shows that the industry giants with the best safety records who lay out the welcome mat when OSHA calls, get hit the hardest. They get the multi-million dollar OSHA fines. The inspectors like to work where they feel welcome and they believe that big business deserves to be hit big.

B.   PREPARATION

Most employers will never receive an OSHA inspection, nor will most taxpayers be audited by the IRS. But you could be one of the unlucky ones, so you should prepare accordingly.

There are literally thousands of OSHA regulations. It's impossible to come up with an accurate count because many of them incorporate by reference privately-developed standards that have never been printed in the Federal Register and are not readily accessible. No one has ever read them all.

Many OSHA standards are so vague and ambiguous that they can be interpreted any way an OSHA inspector wants to interpret them. Some inspectors boast that they can find OSHA violations whenever and wherever they choose. No person with any experience in this field will dispute that.

Consequently, there is no way to assure yourself that you have achieved such a state of OSHA compliance that you will not be cited. But there's a lot you can do to keep the adverse consequences of an OSHA inspection to a minimum.

Over the past 10 years, OSHA has increasingly focused upon paperwork requirements. There are 138 OSHA regulations that require employers to adopt written programs specifying the means and methods the employer has adopted to satisfy particular OSHA standards. Another 366 OSHA standards require that records be kept, documents maintained, notices be provided, or warnings or instructions be given. Some 400 others require the employer to provide instruction and/or to train employees in the safety and health aspects of their jobs, or to limit work assignments to employees who are "certified", "competent", "designated," or "qualified".

Those are by far the most-cited standards on OSHA's annual hit list. They are very popular with the inspectors. They don't require extensive knowledge of the employer's complex work processes. The inspector doesn't even have to get his hands dirty through a visit to the factory floor or the places where construction work is ongoing.

Those paperwork requirements are buried within the fine print and arcane language of OSHA's vast regulatory tomes. There is, however, a book, "The OSHA 500", that lists all of them in a reader-friendly format. It is published by Moran Associates, Inc., and is updated whenever there is a change in the regulations.

Few employers are aware of all of OSHA's paperwork requirements. The inspectors know that. They are aware that they can increase their violation totals by simply asking the employer to produce his OSHA-required paperwork for examination.

Those questions are regularly asked within minutes after the inspection's opening conference begins.

Improper recordkeeping and the absence of required records or documents are a fertile source of OSHA violations. So too, are company papers and documents that, though not required, can be used to show that the company hasn't fully implemented every suggestion that could conceivably improve employee working conditions. Millions of different documents fall within this category. Included are safety audits, minutes of meetings, capital improvement requests, doctor's reports, memoranda of subordinate company officials, even sales solicitations.

For example, a salesman advises you of a new upgrade for equipment that you use. He provides you with a flyer or prospectus. You don't buy, but you keep the information in your files. Later on, an accident happens and OSHA inspects. The inspector learns that you knew of the availability of the upgrade but didn't obtain it. He concludes that it could have prevented the accident. You are cited for knowingly and willfully exposing your employees to hazard by your indifference to the prospective upgrade.

The same scenario applies when the "smoking gun" is contained in safety-meeting minutes, suggestions made by subordinate supervisors, unions, employees, or contained in books or letters, or what-have-you.

OSHA will subpoena your records in an effort to locate documents of that kind whenever there is an accident or industrial calamity. There have even been cases where anonymous tips have caused OSHA to issue subpoenas for literal truck loads of company documents.

Consequently, always include a written response to documents of that kind if you keep them in your files. Explain why the suggestion is impractical, unnecessary or not feasible. Failure to do so can open you up to a "careless indifference" charge.

The foregoing should not be taken as an argument in favor of paper-shredding. You should always retain the OSHA-required records and written programs that have been referenced above, and you should keep records of everything you do -- and every dollar you spend -- on employee health and safety. It will come in handy if you're charged with a willful OSHA violation, or indicted for a criminal offense based upon a workplace event.

Keep records of all employee training and instruction sessions and the issuance of safety rules and equipment to employees. Get signed receipts. Get acknowledgements from job applicants that they are qualified to perform the job they seek. Never forget that a fatality may someday force you to defend yourself in court against charges that the deceased was assigned to work for which he was not qualified or equipped. Some employers have served jail time on such charges.

The vast majority of OSHA violations are based upon information provided to OSHA by **the cited employer**. If he wasn't told, the inspector would never know, for example, how a particular machine works, how frequently its operated, the chemical composition of particular substances, the amount used in work processes, the work schedules of particular employees, and similar things that are needed to show noncompliance with various OSHA standards. Employers should be aware of that phenomena and decide how much and what kind of information they will voluntarily share with OSHA, as well as the circumstances under which it will be disclosed.

Each employer should also adopt a policy on OSHA inspections. It should cover, at a minimum, whether or not warrantless OSHA searches will be permitted and who will represent the company during the on-site inspection.

An OSHA inspection can have as devastating an effect upon a business as a fire, flood or similar disaster. Prepare yourself accordingly. Know the different consequences of OSHA inspection without a warrant and those conducted with a warrant, then adopt a company policy on the issue.

Above all, have a trained, designated representative at each workplace whose job is to represent the company's interest during an OSHA inspection. That person should be thoroughly acquainted with the matters discussed above as well as those that follow. That representative should carefully consider the following suggestions on "Responding to the Inspection."

C.   RESPONDING TO THE INSPECTION

1. No one in management except the company's designated representative should engage in any substantive discussion with an OSHA inspector. When the inspector arrives, have him wait until that representative arrives. Everything that follows are suggestions for your company representative.

2. Greet the inspector politely. Look at his credentials, copy down his name and address, or get a copy of his business card. Ask him **why** he wants to inspect your plant and **why** you were selected for inspection. Make notes of the answers. Save them for future reference. This is important. A number of employers have learned, to their chagrin, that the reasons the inspector gives in person while he is at their plant vary from those given later.

3. If your company has a policy against warrantless searches, ask him if he has a warrant. If you get a negative response, advise the inspector of company policy. Don't give any reason why the policy exists. Simply state that it's contrary to company policy to let him in. Ordinarily he will leave without further discussion. If, however, he asks for additional information, do not provide it. Tell him that it is contrary to the company's policy to provide the information in the absence of a warrant.

4. If the inspector shows up with a warrant, get a copy of it. Read it carefully. Keep it with you throughout the entire inspection. It will state the time limits and the ground rules for the inspection. You'll need it if issues arise over what OSHA is authorized to do. If the warrant refers to any other document, insist upon a copy of that document too.

5. If your company has no policy against warrantless searches (or No. 4 above has occurred), allow the inspection to begin. Invite the inspector in to your office or some similar place where the ground rules for the inspection can be discussed.

6. Treat the inspector with respect, but don't grovel or cringe. Keep in mind that it's your business. He is your guest. Neither of you can dictate to the other but, as in all interpersonal relationships, one person's desires may prevail over the other's. Try to have **your** wishes prevail if possible. For example, express your own views as to where and when various phases of the on-site inspection will be conducted. He may surprise you by honoring your wishes. If you do **not** express yourself on those matters, of course, **his** views will prevail.

7. All OSHA inspections begin with an "opening conference." The inspector has a "check list" of topics to ask you. It will take about an hour to cover them. He will ask you questions about your business similar to those that might be asked by a census taker. For example: What kind of business is this? How many employees? Is it a corporation? Who owns it? Do you operate in interstate commerce? You shouldn't have any problem with those questions but, always keep in mind, you have **UNLIMITED TIME OUTS**. Use them to consult with others whenever a potential problem arises.

The opening conference will also cover the following 14 matters.

a. **Purpose.** The purpose and scope of the inspection, i.e., to investigate an accident or complaint or to conduct a "programmed" wall-to-wall inspection of the entire facility. The OSHA inspector will tell you that. He will also state whether it is a "safety" inspection or a "health" inspection. The latter type of inspection is conducted by an OSHA industrial hygienist and usually includes air sampling to determine if there are any airborne toxic contaminants, and/or measurement of noise levels to see if OSHA noise limitations are being observed. If you want more information than the inspector provides, speak up. Ask anything you please. You are entitled to be provided with full and complete information.

b. **Labor Union?** Whether your employees are represented by a union. If so, he will request a meeting with the shop steward or similar union representative. He will want to include the union in the opening conference. You can decline that request. If so, OSHA will generally hold a separate opening conference with the union.

c. **Escort.** He will advise you that both you and an employee representative are entitled to accompany him throughout the physical inspection of your business. You should always do so. If one of your employees (the shop steward, for example) also chooses to do so, you have the right under the OSH Act to require that the employee do it on his own time. Most on-site OSHA inspections are conducted <u>without</u> employee or union representation.

d. **Injury Log.** Your log of employee injuries and illnesses (and perhaps other records). OSHA will want to look them over to see the kinds of injuries and illnesses listed, their frequency, and whether or not there is a pattern there (such as eye injuries, falls, etc., or a particular job function that is connected to each injury). The OSHA inspector does that so he'll know where to concentrate his attention when he begins the physical inspection. The inspector may also decide to calculate your rate of injury/illness. If so, he'll ask you for an estimate of the hours worked by employees over a year's time. He may also want to verify the accuracy of your injury logs. If so, he'll ask for other records to compare them with.

e. **Handouts.** The inspector will give you various OSHA pamphlets and informational materials, and tell you how to get additional copies.

f. **Sales Pitch.** He will usually give a brief promotional pitch, such as OSHA's goal of reducing worker injuries and illnesses and its desire to help you achieve that objective. You should treat that the same way you treat other sales pitches.

g. **Other Employers?** The inspector will ask if there are people working on the premises who are not employed by your company. If so, he will want to know who they are, what they do, and who they work for. He may want to include other such employers in the opening conference.

h. **Trade secrets.** He will ask if you wish to identify certain matters as trade secrets. If you do, he will promise to keep them confidential -- but he *will* inspect them.

i. **Hazard Communication.** He will ask for a copy of your written hazard communication program together with the list of chemicals that are present on the premises. If you don't have one, you are probably in violation of the OSHA hazard communication standard. Virtually every business in the country is required to have their own written hazard communication program.

j. **Lockout/Tagout.** If you are covered by the OSHA lockout/tagout standard (and if you have any machines or similar kinds of equipment, you probably are), he will ask to see your written program and he may question you about the details of your lockout/tagout procedures.

k. **Other Records.** He may ask you for equipment inspection and maintenance records, certification records, medical surveillance records, results of monitoring conducted (or measurements made) of workplace noise or air contaminants, safety committee minutes, checklists, records of audits or inspections made by insurance companies, consultants or others. You do **not** have to provide him with everything he requests. When in doubt, call a "time out", go to the telephone and call for advice from your attorney or some knowledgeable source that you trust. Be especially careful about records of internal safety audits or recommendations for improvements in your machines or equipment.

l. **Posting.** He will want to see if you have the OSHA poster ("Job Safety and Health Protection") on display. It is required to be posted at every place where people work. He may also inquire about the posting of injury/illness log summaries (required only during the month of February each year).

m. **Safety Programs.** He will ask to see any written safety programs you have so he can check into whether they are being observed during his physical inspection of the workplace.

n. **Photographs and Videotapes.** He will tell you of his intention to use one or both of them during his physical inspection and he may ask if you object. If you do, he'll take a "Time Out" and call back to his office for guidance.

You must be prepared to deal with those 14 "opening conference" subject matters before OSHA shows up. If you are uncomfortable with any of them, get help now.

8. You are not required to either give the OSHA inspector copies of, or let him look at, everything he asks about. He is making a request. You have a right to ask why. You also have a right to call "time out" so you can make a telephone call or consult with others on how you should respond to his requests. You also have the right to say "no".

9. After the opening conference ends, the "walkaround" inspection (or "plant tour") begins. The inspector will look over the workplace, see what work is being done and observe the working conditions and practices. That will generally be followed by a more intensive investigation of things he thinks might constitute OSHA violations. He will probably want to question employees, foremen and supervisors; take photographs; make measurements; do air sampling and things of that nature. You should do likewise. Stay with him at all times. Don't let anyone else take your place. Never let him out of your sight. Take your own notes and photographs. Make your own measurements. Take control. Remember, its **your** business. Act accordingly. If ever unsure, or in doubt about anything, call a "time out" and seek advice.

10. The inspector will try to get you (and perhaps others in management) to explain what particular employees are doing, how certain machines work, what substances and materials are being used, and how much. It's not simply curiosity. His

objective at all times will be to get you to admit to violations of OSHA requirements. He will watch the employees at work to see if they are exposed to -- or have access to -- conditions or practices that he considers to be hazardous. That process may take a few hours, a few days, or weeks or months. It depends upon how long he is permitted to stay, what conditions exist, or what he learns from you, your employees, or other company people. Remember that, unless he is there pursuant to a warrant, he is there with **your consent**. You can withdraw that consent at any time.

11. He will constantly take notes -- on what he sees and hears and what is said to him. Those notes will be the basis of his testimony against you in the event that you dispute the resulting citations and penalties.

12. Some inspectors have been known to give orders to employers or employees. They have no such authority. OSHA can only act through properly-issued written citations and orders. If the inspector exceeds his authority, you may want to remind him of those limitations. If a friendly reminder doesn't work, you may have to

telephone the OSHA Area Director of the office to which the inspector is assigned.

That's one reason why you obtained a copy of his business card when he first arrived.

13. He may point out things that he believes to be OSHA violations. If you agree that they are, you will almost surely be cited and fined. That will happen even if you correct them while he watches. His own job performance evaluation depends upon the issuance of citations and fines -- not upon changes you agree to make in your business.

14. Keep in mind at all times that what you say can -- and will -- be used against you. The vast majority of OSHA violations are based on what the **employer** said to the inspector -- not what **employees** said -- or what the inspector **saw** during the inspection. Don't engage in idle talk or chit-chat. Inspectors will sometimes take things out of context and use it against you.

15. Don't be reluctant to disagree with the inspector whenever he expresses an opinion that a particular condition or practice constitutes an OSHA violation. He is not infallible on any subject. Remember, for a violation to exist there **must** be a reasonable **probability** -- not a possibility -- that the condition or practice could result in injury or illness to an employee. OSHA is **not** a building code. Its **only** purpose is to prevent **employee** injury and illness.

16.   Never automatically accept an OSHA inspector's advice or instruction on anything. Many who have done so have lived to regret it. Few OSHA inspectors are experts on all OSHA requirements. It would be foolhardy to accept an inspector's word on what OSHA requires without first obtaining verification from a source you know to be reliable and well-informed.

17. Do not conduct any demonstrations for OSHA's benefit. In an attempt to gather evidence showing a certain nonoperating machine to be in violation, OSHA may ask you to start it up so they can see it in operation. Don't do it! Thousands of convictions have been based on nothing more than an employer's willingness to conduct just such a demonstration. You don't have to do it. When you get a request to conduct a demonstration for OSHA's benefit -- and this happens often -- remember that it is simply that: a request. A simple "no" from you is all you need to handle this potential problem. It is no different from a request from the inspector to borrow your car. You aren't obligated to do so and it can only work to your disadvantage if you do.

18. The inspector may want to interview some of your employees privately. That is permissible if it doesn't interfere with the employee's work. If it can be done at the employee's workstation at a time when he is not otherwise occupied, stand within view but outside of earshot. Make a note of who the inspector speaks with, as well as the beginning and ending time of the conversation. If the interview cannot be conducted at the employee's workstation without interfering with his work, do not permit the interview at that time. Tell the inspector he is free to obtain the employee's name and telephone number (if the employee will give it) and that he can arrange to conduct the interview at

some other time or place. Keep in mind that the OSH Act guarantees you the right to accompany the inspector "during the physical inspection" of your workplace -- not just some of the time but **all** of the time. This is an important right which you should not allow to be abused. Stay with the inspector during every minute he is on your premises. Remember he is your guest - treat him as such and make sure he never abuses his status as a guest.

19. OSHA inspectors sometimes will want to use your employees as test-subjects or experimentees in order to monitor noise levels or atmospheric conditions in your workplace. To do so, OSHA will not only want to attach measuring devices to their clothing or bodies, OSHA will also want to control the employees' movement and activities during the time such devices are attached in order to insure the integrity of the test results. OSHA does *not* have the right to do this if the subject employee objects or if it will interfere with work operations. The inspector, however, won't always advise employees of their right to decline. If he doesn't, you should. It is also important to remember that OSHA must show that your employees are exposed to allegedly offensive noise or atmospheric conditions for a sufficient number of hours or days. Otherwise, it won't constitute a violation. Consequently, answering such seemingly innocuous

questions as "What hours do employees work in this shop?" or "How often do you use this substance?" could supply OSHA with a vital element of proof in such cases.

20. Never give estimates. You may be providing false information to OSHA, a criminal offense, or you may be thereby wrongfully convicting your company of an OSHA violation. If you don't have accurate and reliable information when asked, say so. But make sure the inspector understands that the information may exist elsewhere. You can tell him that the company would not have any need for the various experts it employs if you knew everything.

21. If the inspector wants to interview another management representative -- and you decide to permit it -- you should also be present during that interview. Sometimes OSHA will try to get contradictory statements from different employer representatives. That can be trouble for your company. Try to prevent it if you can.

22. Make notes of everything that happens -- and when it happens. Many such notes have proved to be worthless because they did not include calendar dates and times of day. Make the same measurements as the inspector. Take the same photographs. If he uses a video camera, do likewise. Get someone to help you with the camera or with your note-taking if needed. A video of the inspector in action can be very helpful. Some inspectors have shown to be less inclined to use videotape when they are aware that they are simultaneously being videotaped. The evidence you gather in this manner will be helpful when defending against a subsequent OSHA citation. You can be sure that the inspector will testify at that trial as to the accuracy of his notes, the preciseness of his measurements, and the fact that his photographs are truly and fully representative of the situations that existed in your plant during the inspection. If you disagree, you can testify to the contrary and maybe you'll convince the judge. But your testimony would be strengthened if you had made the same measurements, photographs, and notes as did the inspector or had a video of the inspector in action. You have the same rights as the inspector and you would be well advised to exercise them. Some employer representatives bring along a stenographer or a cameraman, or carry a portable dictating machine or a tape recorder during the inspection to make their note-taking easier. That is perfectly permissible.

23. While accompanying the inspector, don't hesitate to ask him questions. Indeed, it can be very important to you. Make a record of the answers you receive. Find out why he is making that measurement, what he thinks is shown by the picture he took, what he is writing in his notes. Many times OSHA inspectors will take a photograph in such a way that it will show an apparent violation (a "blocked exit," for example). Your photograph should be taken of the entire scene in order to show the true condition. Don't be bashful. Ask the inspector about his pertinent background, knowledge and experience whenever he shows particular interest in a machine, apparatus, process, or such. He may not answer you fully, but keep track of what he says. It may be invaluable to you if a citation issues. Frequently he will concede that he has never previously seen the type of machine he is claiming to be in violation. Keep a record of any such statement because it is unlikely that he will make such an admission later. If a citation issues and there is a trial, he might then testify that he is one of the world's leading experts in that type of machine.

24. At the end of *each* inspection day go over your notes and measurements. Make sure they are complete and accurate. Have them typed if possible. Be sure to

record the date you made them -- a simple matter which many people overlook. Be certain also to record the name of the OSHA inspector (or inspectors) and, if you are recording conversations which took place, be sure to specify who said what. That includes you -- if you said anything to the inspector. When your photographs are developed, label each of them by what they depict and the date and time they were taken. Once you have done it, carefully retain all of it (including your original notes, measurements and tapes, even when you later have them typed). That will be important to you if you are cited. Even if you aren't cited, this material can be helpful in your efforts to understand and comply with OSHA requirements. Should you again be inspected at some future date, it could be very important on that occasion too.

25. Frequently, the inspector will give you copies of publications or other documents. Don't refuse them even if you already have copies. Keep all of them. Write on each one the date it was given to you and the name of the person who gave it to you. There have been occasions when employers have been given out-of-date publications by OSHA inspectors -- not the documents an OSHA inspector may later claim that the

employer was given. Because such matters may have importance to the resolution of

contested OSHA citations, you should take the few minutes necessary to make the

identifying notes on the publications you receive and retain the same for future reference.

26. Be particularly careful about providing OSHA with company documents.

You do not have to give OSHA any records unless a warrant requires you do so. In such

cases, the warrant will ordinarily authorize OSHA to *inspect* certain records. That is

exactly what you should permit when faced with such a warrant. Let OSHA look at the

particular record. Do not give them a copy. If the OSHA inspector wants to copy the

record down by hand, let him do so. It is not to your advantage, however, to give him an

extra copy or let him use your photocopying equipment. Should you do so, prepare to

defend yourself against a citation based upon the record you supplied. OSHA can

subpoena records if you don't voluntarily supply them. That is generally better for you

than disclosure upon request because (a) the inspector may decide it's not worth his while

to go through the subpoena process, or (b) you can always negotiate the terms of a

subpoena once it has issued -- and you can sometimes get all or part of it set aside.

27. Resist the temptation to voluntarily provide OSHA with safety rules, audits, studies, or other things that in your opinion reflect favorably upon your company. In many cases, that material can be used **against** you. For example, in 1975 the Forging Industry Association retained a university-based team of noise experts to study ways of controlling noise in forge shops. A lengthy study produced a comprehensive report concluding that there were no feasible engineering controls to reduce forge hammer noise to within OSHA limits. Interspersed within the report were a few suggestions for lowering noise from other sources. OSHA was supplied with copies of the report because the forging industry thought it would deter OSHA from issuing noise citations. OSHA, however, used it to cite forge shops for **willful** violations because the forging companies had not done those things the report had simply suggested. Similar things happen all the time. That's one of the reasons why supplying OSHA with documents can cause you trouble.

28. When the plant-tour phase of the inspection has been completed, OSHA procedures call for a "closing conference" with the employer. This is *not* to be thought of by the employer as a friendly chat or a chance for you to talk the inspector out of a

citation. It is the time when the inspector discusses his findings and looks for admissions

from the employer of facts which will help OSHA prove their violation charges. Most

admissions against interest are made by employers during the closing conference. You

can be sure that the inspector will be looking for such statements and will take note of

them. Indeed, the *OSHA Field Operations Manual* **requires** him to do so. It tells the

inspector that in filling out his report of the inspection, he should show any supportive

information from the closing conference substantiating the employer's general attitude,

any general admission of violations, and any agreement about abatement dates. It is not

uncommon, for example, for an employer to respond to the inspector's description of a

condition observed during the inspection by saying: "Well, we've been trying to get that

changed but can't do it." That statement is good evidence both that a substandard

condition exists *and* that you, the employer, have been aware of its existence for some

time. That statement will cost your employer plenty. You might be better off to bite your

tongue.

29. The inspector is supposed to use the closing conference to advise you of

"apparent violations" he has observed during his inspection. If he uses the word

"violation," it is never to your advantage to agree with him. Semantics are very important

here. You may concede that a particular "condition" or "practice" exists -- but that doesn't necessarily establish that it constitutes an OSHA "violation" (there's lots of law

on this point). However, when the inspector states that this or that condition or practice

is an apparent violation, don't agree, but do ask him (as to each) why **he** thinks so -- and

try to find out what, in his opinion, would constitute a feasible means of abatement. Get

as many details on this point as you can, but do not argue with him about his response.

For example, if he says "I don't know," make a note of it and move on to the next

question. If he states that you can correct the excessive noise in your machine shop by

buying a can of anti-noise spray at the corner drug store and spraying the machine with it

every Tuesday at noon, don't tell him he's crazy. You'll have plenty of time for argument

later (after you have contested the resulting citation).

30.   The inspection is the time for acquiring information. Get as much of it as

you can -- no matter how irresponsible you may think that information to be. Be sure,

however, to record the inspector's responses fully. You can do so either by a tape

recorder, your own note-taking, or by having a stenographer present. Perhaps you should

not use the word "feasible" during your questioning, since it tends to raise red flags in the

minds of OSHA people. What you want answered are the details of what he thinks you

ought to do in order that the condition or practice he referred to will be corrected or prevented from happening in the future. You want all possible details on the means and

methods of abatement, what disciplinary, instruction or employee training program he

believes that you should have (including all the details thereof), a full list of all

equipment and materials needed, how much they will cost, where they can be acquired,

what result they will accomplish, and precisely where they must be installed. Ask those

questions. Also ask what safety and health benefits your employees can expect if you do

what the inspector wants you to do. Have him explain the "how" and "why" of each

position he takes.

31. Be careful in responding to this question which he will be almost certain to

ask: "How much time will you need for abatement?" An unexplained time-estimate by

you can be interpreted as both an admission that there **is** a violation **and** that your

company will be able to abate it within the time period you state. If you don't think the

condition mentioned is a violation, say so. Tell him there is no need to abate -- or there is

no feasible way to do so. Don't worry if he tells you in response that an arbitrarily brief

abatement period will be assigned. You'll receive an automatic extension of any such

time when you contest the citation.

32. You are not required to have a closing conference and if you lack the courage or the skills necessary to avoid unintended admission against your interests, you would probably be wise to tell the inspector that you do not want any closing conference. Don't be fearful that you'll miss out on something if you skip the closing conference. There will be plenty of later opportunities to ask questions and conduct negotiations with OSHA. They will come **after** citations have been issued and you have contested them. That is usually a good time to get information because you don't have to worry about what you say being used against you. There is also another alternative. You can request that the closing conference be conducted by telephone. OSHA will generally grant such a request.

33. Once the inspection is concluded either with or without a closing conference, the inspector returns to his office to do his "write-up" on the inspection. That may take days or weeks. Occasionally, during that process, he may telephone you with a request for additional information or documents. You have the same rights (mentioned above) to say "no" to his request or to delay a response until you have conferred with your own attorney or any one else you rely upon.

34. The inspector will prepare the citations and proposed penalties after returning to his office. They will be sent to you by certified mail (or personally delivered) a few weeks or a few months after the inspection ends (except that in California, handwritten OSHA citations are sometimes given to the employer before the inspector leaves the worksite). Remember that OSHA must act by means of written citations. An inspector will occasionally forget that fact, and will tell you (or your employees) that this or that must be done. Those verbal orders are **not** enforceable and you are not under any legal obligation to observe them. A wise policy to follow is: Get it in writing. That applies to every order he gives, or advice or instruction that he offers. If he tells you, it's in the regulations or contained in a particular publication, have him list the **specific** section, subsection, paragraph, page, and sentence he is relying upon. Don't accept generalities.

35. The citations and penalty proposals **will** be in writing. They will be typed upon a government form and will be accompanied by a 2-page form-letter from the OSHA Area Director, a booklet explaining your rights and responsibilities, and perhaps some other OSHA hand-outs. You will have 15 working days (approximately 3 calendar weeks) to contest the citations. You should **always** do so. It's a very simple process -- a one-sentence letter will do the trick (a suggested letter is printed at a later point in this book). Don't worry about a hearing. You'll have at least two months after you contest to negotiate a settlement. OSHA will be anxious to have that occur. They don't like hearings either. Keep in mind that citations that are **not** contested are the moral equivalent of guilty pleas. That can come back to haunt you. There are ways to settle contested citations that can avoid or ameliorate potentially dire consequences in the future. Consult with people who are knowledgeable on such matters and follow their advice. It could be very important to your company in defending against law suits, criminal indictments and future OSHA confrontations.

36.   Frequently, an OSHA inspector will state that a "hazard" exists. If you do not agree with his assertion that a particular condition or practice is hazardous, don't be bashful. Tell him he's wrong whenever you think he is. If you don't, you may create additional problems for yourself. Here's an example:

> During the closing conference, the inspector stated that he had found nearly a hundred OSHA violations during his 3-day inspection. He asked why there was so many. The company's safety director replied: *"I'm only one person. I can't do it all."* The inspector treated that reply as an admission that the company knew about the alleged violations. He therefore cited all of them as "willful" with a penalty of $300,000 rather than the $30,000 fine he was considering.

That's an example of how your own words can hurt. Those "willful" citations probably would not have issued if the safety director had replied by asking the inspector for an itemization of his claim, or had simply responded: "That's your opinion."

37.   During the closing conference, the inspector will be guided by a "closing conference checklist" of matters to be covered. A copy of a checklist that was used during an actual OSHA inspection follows:

CSHO # _____ Report # *106212491*

☑ Act (Public Law 91-596)

☑ Employee Rights & Responsibilities
Following An OSHA Inspection.

____ Asbestos

____ Recordkeeping Requirment/Forms

____ What Every Employer Needs to Know
about OSHA Recordkeeping

☑ OSHA Poster

____ Hand and Power Tools

____ Ground Fault Protection

____ Seat Belt Flyer

☑ All About OSHA

____ Training Requirements

____ Respiratory Protection

____ Excavation & Trenching

____ Teaching Safety & Health

☑ Electrical Standards for Const.

____ Personal Protective Equipment

☑ Hazard Communication Guidelines

☑ Consultation Services for
the Employer 7(c)(1)

____ OSHA Publications & Training Mats.

____ Material Handling and Storage

____ Other _____

ALL ALLEGED VIOLATIONS WERE DISCUSSED AND REASONS GIVEN WHY THEY DID NOT COMPLY
WITH THE STANDARDS. THE FOLLOWING OSHA RULES AND PROCEDURES WERE DISCUSSED:

☑ Citations

☑ Explain Serious vs. Other

☑ Penalty Assessment

☑ Posting

☑ Informal Conference Procedures

☑ Contest Procedures

☑ Multi-Employer Policy

☑ Abatement and Extensions

☑ Abatement Letter & F/U Inspections

☑ Referrals

☑ 11(c)(1) Restrictions

☑ Consultative Services

☑ Hazard Communication

____ Lock-Out Procedure

III. Recordkeeping Adequate? *N/A*   Deficiencies _____

____ '85 ____ '86 ____ '87 ____ '88 ____ '89 ____ '90 ____ '91

OSHA 200 posted: *YES*

IV. Complete inspection: Yes ☑   No ____
If no, specify areas
covered or not covered: _____

If F/U, specify items covered: _____

V. Penalty Adjustment
Size: 0-10, (11-25,) 26-60, 60-100, greater than 100   *60*

Good Faith: _____   *0*

History: _____   *10*

Total   *70*

6. **Citations and Penalties**

OSHA citations and penalty proposals ordinarily arrive several weeks after the inspector has completed his on-site inspection and departed from your premises. They will arrive in a package that will include routine boiler-plate information that may be of little or no interest to you. The citation, however, is important. It will be the single most important part of that package. OSHA people call it the "OSHA-2" form. It is entitled "Citation and Notification of Penalty". An example follows on the next page of this book.

U.S. Department of Labor
Occupational Safety and Health Administration

Citation and Notification of Penalty
Savannah Area Office
US Department of Labor - OSHA
500 Bull Street - Suite 500
Savannah, GA 31401

3. Issuance Date	4. Inspection Number
12/12/91	105212491
5. Reporting ID	6. CSHO ID
0410940	ME232
7. Optional Report No.	8. Page No.
017	1 of 2

The violation(s) described in this Citation are alleged to have occurred on or about the day the inspection was made unless otherwise indicated within the description given below.

10. Inspection Date(s):

9/25/91 - 10/17/91

1. Type of Violation(s)	2. Citation Number
Serious	01

9. To: HYPOTHETICAL CO., INC.
P.O. Box 202
Midway, GA 31443

11. Inspection Site:

100 General McIntosh Blvd.
Savannah, GA 31401

Penalties Are Due Within 15 Days of Receipt of This Notification Unless Contested (See enclosed Booklet)

This Section May Be Detached Before Posting

THE LAW REQUIRES that a copy of this Citation be posted immediately in a prominent place at or near the location of violation(s) cited below. The Citation must remain posted until the violations cited below have been abated, or for 3 working days (excluding weekends and Federal holidays), whichever is longer.
This Citation describes violations of the Occupational Safety and Health Act of 1970. The penalty(ies) listed below are based on these violations. You must abate the violations referred to in this Citation by the dates listed below and pay the penalties proposed, unless within 15 working days (excluding weekends and Federal holidays) from your receipt of this Citation and penalty you mail a notice of contest to the U.S. Department of Labor Area Office at the address shown above. (See the enclosed booklet which outlines your rights and responsibilities and should be read in conjunction with this form.) You are further notified that unless you inform the Area Director in writing that you intend to contest the Citation or proposed penalties within 15 working days after receipt, this Citation and the proposed penalties will become a final order of the Occupational Safety and Health Review Commission and may not be reviewed by any court or agency. Issuance of this Citation does not constitute a finding that a violation of the Act has occurred unless there is a failure to contest as provided for in the Act or, if contested, unless the Citation is affirmed by the Review Commission.

12. Item Number 13. Standard, Regulation or Section of the Act Violated	14. Description	15. Date by Which Violation Must Be Abated	16. Penalty
1 29 CFR 1926.500(d)(1): Open-sided floors or platforms, 6 feet or more above adjacent floor or ground level, were not guarded by a standard railing or the equivalent on all open sides: (a) Second, third, fourth, fifth, sixth, seventh and eighth floors of building under construction - No standard railings and toeboards provided, on or about 9/25/91.		Immediately Upon Receipt	1500.00

17. Area Director		18.
Luis R. Santiago		

NOTICE TO EMPLOYEES — The law gives an employee or his representative the opportunity to object to any abatement date set for a violation if he believes the date to be unreasonable. The contest must be mailed to the U.S. Department of Labor Area Office at the address shown above within 15 working days (excluding weekends and Federal holidays) of the receipt by the employer of this Citation and penalty.

EMPLOYER RIGHTS AND RESPONSIBILITIES — The enclosed booklet outlines employer rights and responsibilities and should be read in conjunction with this notification.

EMPLOYER DISCRIMINATION UNLAWFUL — The law prohibits discrimination by an employer against an employee for filing a complaint or for exercising any rights under this Act. An employee who believes that he has been discriminated against may file a complaint no later than 30 days after the discrimination with the U.S. Department of Labor Area Office at the address shown above.

Total Penalty for This Citation

Make Check or Money Order Payable to "DOL-OSHA"

Indicate Inspection Number on Remittance

Familiarize yourself with the "Citation and Notification of Penalty." Notice for example that it contains little numbers -- 1 through 18 -- that generally follow a left to right arrangement beginning just above where your company name is typed. Numbers 1 through 11 are contained in the upper half of the page. The remaining numbers (12 to 18) are found on the lower part of the page. Your company name will be typed in entry number "9", after the word: "To:" (In the example on the preceeding page, the company name is listed as "HYPOTHETICAL CO., INC.,").

If you are lucky, there will be only one of them. In most cases, however, there are many of them. (One sorrow-stricken company reported that it received over 1,000 pages of OSHA citations when the inspection was over).

There are at least 4 things on the "Citation and Notification of Penalty" that should attract your attention:

First. The "Description" of the alleged violations (i.e., the cited condition or practice). I will cover several lines that will be typed in approximately the middle of the page, just <u>below</u> the printed entry: "14. Description."

Second. The "TYPE" of violation (willful? serious?). That's numbered "1", just above your company name.

Third. The Amount of proposed penalty. It is number "16", on the far right of the page, approximately half way below the top of the page.

Fourth. The Time allowed for abatement. That is number "15," just to the left of number 16.

Ordinarily, the last-mentioned of those four should concern you the least. It will be **automatically** suspended if you contest that citation and, even if you don't contest, OSHA is usually willing to grant a time extension if you can provide some good reason why you need more time.

You will notice that the time allowed for abatement is listed just to the right of the description of each alleged violation in a column labeled:

"15. DATE BY WHICH VIOLATION MUST BE ABATED:"

Sometimes a calendar date will be typed in that column (such as "04/23/94"). Or the word "Immediately" might be typed there. No matter which it is, it will be **automatically** suspended if you file a Notice of Contest within the permitted time-period. See the next section of this book entitled: "Contesting Citations".

There is another important matter on the citation that you would be well advised to understand. That is the time of alleged violation. Look just above the printed entry "11. Insepction Site:" in the upper middle of the page. There you will see the following statement:

"The violation(s) described in this Citation are alleged to have occurred on or about the day the inspection was made unless otherwise indicated within the description given below."

The "day the inspection was made" is usually a period longer than a single day. It appears just to the right of the above-quoted statement under the rpinted entry: "10. Inspection Date(s):" Whatever calendar dates are typed there will be the time when OSHA claims that the alleged violation existed unless a different calendar date is included as part of the "description" (No. 14) of alleged violation. Notice that on the sample citation, there is a different time included in the "description" ("on or about 9/25/91").

Why is the time of alleged violation important? Because, if you contest, OSHA must prove its existence at the particular time specified. Not at an earlier or a later time. If the stated condition did not exist at the time specified, OSHA will not be able to sustain its allegation of violation.

The OSH Act contains a 6-month statute of limitations. Consequently, the time of alleged violation must occur within 6 months of the citation's "Issuance Date." That date appears in entry number "3" near the top of the page.

Most employers will focus upon the amount of penalty (No. 16) and the classification (or "type") of the citation (No. 1).

There will be a separate penalty amount for **each** violation cited. If no penalty is proposed, the entry (No. 16) will be: "$0".

The following is a list of "types of violations" (No. 1) that may be cited, and the "penalties" (No. 16) that may be proposed.

- **Other Than Serious Violation** - A violation that has a direct relationship to job safety and health, but probably would not cause death or serious physical harm. If you are cited for that kind of violation the word "Other" will appear in the box (No. 1) labelled "Type of Violation(s)". A proposed penalty of up to $7,000 for each such violation is possible.

- **Serious Violation** - A violation where there is substantial probability that death or serious physical harm could result from the cited condition or practice and that you knew, or should have known, of the condition constituting the alleged hazard. A penalty of up to $7,000 for each violation can be proposed. The word "Serious" will be typed in Box No. 1 of the Citation if OSHA charges you with this type of violation.

- **Willful Violation** - A violation that the employer knowingly commits or commits with plain indifference to the law. In other words, you were aware that a hazardous condition existed. You also knew that there was an OSHA requirement that prohibited it; however, you made no reasonable effort to eliminate it. Penalties of up to $70,000 can be proposed for each willful violation, with a minimum penalty of $5,000 for each such violation. When this type of violation is alleged, the word *"Willful"* will be typed in Box 1.

- **Repeat Violation** - If your company has previously been found in violation of any OSHA standard, regulation, rule, or order and, upon reinspection, a substantially similar violation exists, OSHA can cite you for a "repeat" violation. That can bring a fine of up to $70,000 for each such violation. To be the basis of a repeat citation, the original citation must be final. An earlier citation that you have contested may not serve as the basis for a subsequent repeat citation if the case has not yet resulted in a final order. The word "Repeat" will be typed in Box No. 1 if OSHA charges you with this type of violation.

- **Failure to Correct Prior Violation** - Failure to correct a prior violation may bring a civil penalty of up to $7,000 for each day the violation continues beyond the prescribed abatement date. That kind of violation will appear on a form that looks like a citation but isn't. It will have the words "FAILURE TO ABATE" stamped in the middle of the page where the description of the violation is typed (No. 14). The difference between a "failure to abate" and a "repeat" is that a repeat violation only has to be similar to a prior violation. A failure-to-abate must be the same violation that was previously established.

Those are the 5 different *"Type(s) of Violation"* for which you can be cited by OSHA. Citation and penalty procedures may differ somewhat in states with their own occupational safety and health programs.

Do **not** think of an OSHA citation as an "order" or a "violation." If you exercise your right to contest, both the "time for abatement" and the "penalty" is automatically suspended. No citation can be regarded as an order to make any changes in your workplace condition or pratice, or an order to pay any penalty **unless** you fail to respond to it within the alloted time period.

It is also important to remember that:

> An OSHA inspector has no authority to shut down your business or order you to stop anything or make any changes. A <u>court</u> can do that if OSHA convinces the judge that an "imminent danger" situation exists. That is a very rare occurrence.

OSHA can only issue citations that **allege** violations and **propose** penalties. Those allegations and proposals can only become violations in one of two ways:

- You don't contest them.
- You do contest them, and after a hearing, a judge affirms them.

There is no reason for panic when you receive your OSHA citations. You have a lot of rights. If you contest, you'll probably get off with less. There are also settlement options available to you. They will be discussed in the next few pages of this book.

4. Informal Conference and Settlement

When you receive the citations, you have the right to contest them. It's **very easy** to do. How that is done will be covered in the next section: "Contesting Citations".

You have 15 "working" days (or approximately 3 calendar weeks) to file what is known as a "Notice of Contest". It is simply a letter that you write to OSHA. You can copy from the example that appears in the next section of this book.

During that 3-week time period, OSHA will want you to participate in an "informal conference" with the supervisor of the OSHA inspector that conducted the inspection (the Area Director or his designee). One of the form letters that you'll receive in the same package as the citations will tell you about it.

One of the reasons why OSHA tries to get you to participate in such a conference is to dissuade you from exercising your right to contest the citations. There is no provision in the OSH Act for a post-citation conference. It's an optional, OSHA-created process that some experienced people view with suspicion.

The mere fact that OSHA provides that kind of a conference procedure -- and actively tries to persuade you to participate in it -- is an implicit concession that it has some doubts about the citations it issued to you. Look at it this way:

> If OSHA truly believed that it had arrived at the correct citation and penalty proposals when it issued them -- as it is obligated by law to do -- a simultaneous invitation to the employer to attend a "settlement" conference would make no sense at all.

Nevertheless, it is current OSHA routine to advise each cited employer that he is invited to come to the local OSHA office and confer with the OSHA Area Director. You are free to accept that invitation, to decline it, or to ignore it altogether.

You can also attend the "informal conference" and, if not satisfied with the result, then exercise your right to contest. Many employers have done that. But it must be done quickly. **No one** can give you an extension of the 15 working-day limit for contesting. If you miss that deadline, you're out of luck. The citations and penalty proposals automatically become final and binding when the deadline passes without a Notice of Contest.

Although called an "informal conference", there are few things about it that are informal. The conference is carefully planned by OSHA to the extent that the agenda, purpose, list of participants, and even who will take notes, are specified in various OSHA manuals and directives.

Some cited employers take advantage of the procedure and obtain satisfactory reductions in the penalty proposals and adjustments in the citations. Some come away empty-handed. Most don't bother.

Speed is essential to the holding of the "informal conference" because the 15-working day time period for filing a Notice of Contest *cannot* be extended.

It doesn't really matter if you decide against an informal conference because the opportunity to settle is *always* there for those employers who file their Notice of Contest on time. Indeed, the vast majority of settlements have occurred *after* the Notice of Contest was filed.

Many people experienced in the process believe that the employer's settlement prospects improve with age and time. That belief is based, at least in part, on the fact that once a Department of Labor lawyer takes over the case (which happens just as soon as you file a Notice of Contest), he or she may have some doubts about OSHA's prospects for prevailing in the litigation, and even if not, may not want to go to the considerable trouble of preparing the various pleadings that are required in all contested citation proceedings.

Further details on settling OSHA cases appear in the next section of this book entitled "Contesting Citations".

Clearly, time is on the side of the employer who contests because he probably will not receive another OSHA inspection while the case is pending and he is under **no** obligations as a result of his receipt of the citation during that period of time (except to post a copy at the workplace).

Trial of a contested case often will not occur in the same calendar year that the citation is issued and sometimes it will be a considerable period of time before the trial is scheduled.

The matters discussed above are among several factors that weigh against the quick settlements sought by OSHA in an "informal conference" and in favor of immediate filing of a Notice of Contest, then settlement at a later time.

IV.   CONTESTING CITATIONS

1.   How to Contest

Contesting an OSHA citation is as easy as writing a letter. Indeed, that's all there is to it. It doesn't have to go special delivery, registered mail, or be served by a process-server. A thirty-two cent stamp will do the trick (at least until the next increase in first class mail).

The pages that follow will discuss the pros and cons of contesting the citation and explain the processes and procedures involved.

At the end of that discussion, there will be a summary of what happens at the various stages of a typical contested OSHA case.

Notice of Contest

When you prepare your own Notice of Contest, you should use your own company letterhead stationery. Address it to the OSHA Area Director whose name will be typed at the lower left of each citation page beside the printed entry: "17. AREA DIRECTOR." The address of his office will be typed at the **upper** left of each citation page directly beneath the printed words: "Citation and Notification of Penalty."

You don't have to give any reason for contesting a citation and you shouldn't volunteer a reason. It won't help your case and, under some circumstances, might cause you difficulties at a later time. So, be brief and direct. You can't go wrong if you simply follow the Notice of Contest form printed above. That form does include a reason for contesting the proposed penalty (because it's "too high"). The reason for that is explained later in Part 2.

2. What Happens When You Contest

An OSHA citation is simply an *allegation* that the employer violated one or more OSHA requirements. The monetary "penalty" amount is simply a proposal. Either or both can be contested. Because they are only "allegations" and "proposals", the employer who contests them is **not** filing an "appeal" and that word is not used at all in the process.

The citation's "abatement date" or, as stated on the citation "Date By Which Violation Must Be Abated", is an integral part of the citation. When the citation is contested, so too is the time for abatement. You do **not** even have to **mention** abatement date in the Notice of Contest. The filing of a Notice of Contest **automatically** suspends the time for abatement stated on the citation. You are not required to take **any** abatement action for **any** citation that you contest until after you voluntarily settle the matter or a judge, after a hearing, affirms the contested citation. IF NEITHER OF THOSE EVENTS OCCURS, NO ABATEMENT OBLIGATION EXISTS AT ANY TIME.

Once the OSHA Area Director receives the employer's Notice of Contest he is obligated to "immediately" forward it to the Occupational Safety and Health Review Commission. Thus begins a whole new ballgame. You are now in the **best** position you've been in since the OSHA inspector first appeared at your worksite.

The Review Commission is an independent agency of the U.S. Government. It has the same power as a court (and is often called a court). It is **not** connected in **any** way to the Department of Labor or OSHA.

There are about twenty Administrative Law Judges (ALJs) who are employed by the Commission. They are stationed at various places throughout the country. However, they come to you to hold their hearings because their rules provide that all hearings will be held at the place of alleged violation or as close thereto as possible. The ALJs are independent. They have lifetime terms of office. They are not beholden to anyone. And that independence is reflected in their decisions.

Each of the 21 "state plan" states has similar (but not necessarily identical) judicial officers and procedures.

At the time the OSHA Area Director sends your Notice of Contest to the OSH Review Commission, he will enclose in the same envelope the contested citations and penalty proposals. At that point, a completely different phase in the process begins.

The citation, penalty proposals and your notice of contest become an OSH Review Commission "case". OSHA is one "party" to the case, designated as the "Complainant". You are the other "party" to the case, designated as the "Respondent". The two "parties" stand on **equal** footing before the Commission.

As in most judicial proceedings, the Respondent (your company) is presumed to be innocent of the charges contained in the citation. The burden of proof is on the Complainant (OSHA). If OSHA fails to sustain that burden of proof (and that is <u>not</u> unusual), the contested citation must be vacated.

Two other matters of interest occur at the same time that OSHA turns the case over to the OSH Review Commission.

The OSHA inspection file (including those notes the inspector took during the inspection) is physically transferred to a Department of Labor lawyer who controls all further proceedings for OSHA's side of the case.

OSHA is unlikely to attempt any further inspections of the employer while the case is pending before the Commission - which process frequently takes a year or more to complete.

3. Collateral Considerations

Experience has shown that if you are cited for 10 (or 100 or 1,000) OSHA violations and are especially upset by only 1 or 2 of them, you should nevertheless contest **everything**.

It is permissible to limit a Notice of Contest to the penalty amount only or, if cited for several violations, to only some of them. That is not generally done, however, for a number of reasons including the fact that it can complicate the wording of a Notice of Contest (and OSHA's interpretation thereof) as well as needlessly waste the employer's "bargaining power" in settlement talks that are sure to follow.

There is absolutely no down side to contesting everything. You can withdraw any (or all) matters that you have contested at any time.

> Always state that you contest penalty proposals because they are "too high".
>
> In the past, some have taken the position that when an employer simply states that he "contests" a penalty proposal, it could be because he thinks it was too low and wants it raised.

It is a virtual certainty that, following receipt of the Notice of Contest, an effort will be made by OSHA or its attorney to settle. They are nearly always willing to do one or more of the following in order to avoid the necessity of a trial:

Reduce the total proposed penalty (by as much as half).

Extend the time for abatement.

Compromise on what means and methods will be necessary to accomplish

Reduce the citation's classification (from "willful" to "serious", for "abatement".

Vacate one or more of the contested citations.

Far more than half of all contested cases are settled in that manner without ever going to trial. Few people familiar with such realities would advise an employer to "limit" his contest to only some of the cited violations, or to only the amount penalties.Contesting everything is generally the best policy.When a citation is contested the matter is suspended until it is either:

(1) settled by mutual agreement between OSHA and the employer, or (2) decided by an OSH Review Commission judge following a trial. Therefore, until one of these two events occurs, the employer is under **no** OSHA obligation as the result of the receipt of the contested citation (except to post a copy of the citations).

No abatement of any alleged violation is required and no changes of any kind need to be made in conditions or practices at the worksite while a citation is in litigation. And a citation is "in litigation" from the moment you sign the Notice of Contest and drop it in the mail.

A letter contesting a citation puts everything listed on the citation into contest. There is no reason therefore to separately state that the abatement date is contested.

The penalty proposal, however, comes under a different provision of the Act and requires separate treatment. That is why the typical Notice of Contest contains two separate paragraphs: one for the citations and one for the penalty proposals.

4. Should You Contest?

Is it wise to contest OSHA citations? The answer to that is "yes" unequivocally. There are no disadvantages to contesting. You can **not** receive a penalty greater than that set forth on the citations or be dealt with more severely -- either at the time you contest or in the future. So you can't lose. But you can win. For example:

- You can nearly always obtain a reduction in the amount of proposed penalty.
- You can generally avoid further OSHA inspections while your case is pending.
- There is no abatement obligation until after the proceeding is concluded.
- It can cost as little as the price of a postage stamp.

Compare that with the consequences of not contesting.

The OSH Act makes it quite clear that an uncontested OSHA citation within the 15 working-day limit is final and binding, and cannot thereafter be subject to review by any court or agency. Failure to contest is the moral equivalent of a plea of **GUILTY**. You must abate all cited conditions within the time stated on the citations. You must pay the fine. You are subject to immediate re-inspection and a whole new round of citations and fines. That's not all. There are additional consequences. Read on!

5.   What Happens When You Don't Contest

By failing to contest an OSHA citation, the cited employer thereby pleads guilty as charged. Although few employers realize it, that gives them a "record" as a violator of a law designed to protect employees' lives from on-the-job hazards. It's not a good record to have. To the employer, it may seem innocuous. But to others, it will be viewed somewhat differently.

Others won't look at the **nature** of the particular violations. To them, **every** OSHA violation -- no matter how insignificant you may regard them -- is a black mark on your record.

It is not at all unusual for uncontested OSHA violations to be used by persons or organizations who wish to tarnish a company's otherwise good image. The company's "record" in this regard can be obtained by anyone who requests it. Quoted below are two lines that appeared in a newspaper story about a tragic accident at a shipyard:

"The shipyard, one of the nation's largest repair and construction docks, has a long history of safety violations at OSHA. Documents show that in the past eight years the company has been cited about 235 times for safety violations."

That "long history" resulted from the fact that the company had never contested an OSHA citation in 20 years of being inspected by OSHA. The company policy was to accept all OSHA citations as valid, immediately correct the cited conditions and pay the proposed penalties. They thought that would make them look "good." They learned, however, that it had the exact opposite effect. After the company learned that their cooperative policy was regarded by others as a "bad" record, the policy was changed.

An employee's OSHA "record" may never cause the employer any future difficulties but it will be there.

Government has many faults, but failure to retain records of its enforcement activities is not one of them. Employers cited for OSHA violations in the past ten years can rest assured that records of those violations are carefully preserved and that they will still be retained in the government archives into the next century and beyond. OSHA itself can resort to those records at any time, but they are also subject to release to anyone who asks for them under the Freedom of Information Act.

Your OSHA violation records can be obtained by unions, employees and their estates, other business enterprises, or anyone else who may want to demonstrate that you have not shown appropriate concern for employee safety and health. That is not something that might happen in the future. It is happening right now.

In addition, various government agencies -- and some private companies -- have adopted the practice of refusing to award contracts to, or to do business with, OSHA violators.

If you have OSHA convictions on your record, you can be pretty sure that there is now -- or will be at some time in the future -- someone who will want to use that against you. It may not seem like much to you but some people will think unfavorably of your company if they were to read in the paper -- or be told when on a jury sitting in judgment -- that your company has three, or ten, or more OSHA violations on its record.

Here's an example of what could happen:

A number of employers have been sued by employees who were disabled as a result of their work. Their lawyers used <u>uncontested OSHA citations</u> as ammunition to prove that occupational safety and health problems were called to the employer's attention by those citations and that the employer admitted knowledge of wrongdoing by not even contesting the charges. Some have charged that those citations -- and the fact that the employer did not contest them -- was alone sufficient to establish the employer's "gross negligence." And proof of "gross negligence" is, after all, what is needed in many cases to overcome the barrier in state worker comp laws that bar employee lawsuits against their employer.

Here's another possibility! An accidental fatality occurs. It gets the attention of the local district attorney. In deciding whether or not to charge the company (or its supervisors) with negligent homicide, he looks up their past record of OSHA violations. If it shows uncontested OSHA citations, the District Attorney concludes that the company is indifferent to its worker safety obligations. He decides to proceed with a criminal indictment.

There is an additional problem for publicly-held corporations: potential shareholder derivative suits against corporate managers who allegedly exposed the corporation to the huge losses that criminal prosecution or employee-hazard lawsuits bring. When the door to such a lawsuit is opened because the manager decided to "plead guilty" to OSHA citations by not contesting them, questions may be raised about the manager's fitness.

When a corporation suffers reverses, the search for a scapegoat can be relentless. Often unions, foreign trade or the government itself will conveniently fill that role. But the scapegoat's collar is also fitted around the neck of a business executive from time to time.

6. Settlement of Contested Citations

It is difficult to understand why managers expose themselves to such draconian possibilities when there exists a very simple method that could prevent OSHA citations from becoming a source of future legal problems:

> Contest the citation.
>
> Then settle the case with an agreement that
>
> provides that none of the citation's allegations
>
> may be used in any future civil litigation.

Many enlightened business managers have been following that practice routinely for several years. It is a very simple process. You contest every OSHA citation when received. You later agree to settle the matter without a trial. The agreement includes what lawyers and judges know as "exculpatory" language. OSHA is happy to oblige. OSHA will even draft an agreement for you that contains the necessary language.

A typical agreement of that kind -- one that OSHA itself has agreed to in thousands of cases -- would recite the terms upon which the parties settled the case, then say something like this:

None of the foregoing agreements, statements, stipulations and actions taken by Respondent shall be deemed an admission by Respondent of any of the allegations contained in the citations. Respondent specifically denies each such allegation. The agreements, statements, stipulations and actions recited herein are made solely for the purpose of settling this matter economically and amicably without further litigation and they shall not be used by anyone for any purpose, except for subsequent enforcement proceedings filed by OSHA against the Respondent under the Occupational Safety and Health Act of 1970.

That's just an example. A lot of other things can be said to make it clear that you are **not** admitting to **any** OSHA violation. You are simply agreeing to a settlement because it will cost too much money to go to trial and prove your innocence. Everybody understands that trials are expensive. Few people can legitimately find fault with a company that settles an OSHA case on that basis.

The difference between a settlement on terms like those stated above -- and leaving an OSHA citation uncontested -- is enormous. An **uncontested** citation is at least the moral equivalent of a guilty plea. A citation that is settled with "exculpatory" language is just the opposite. It clearly states that you DENY each alleged OSHA violation and that NO ONE (except OSHA itself) can use the settlement against you FOR ANY PURPOSE.

It would be even better for the employer if the exculpatory language also prevented OSHA from any future use of the settlement agreement. But OSHA won't sign such an agreement because they think it would effectively eliminate their authority to cite the employer for "repeat" violations. OSHA therefore insists that it be excepted from the exculpatory language (that exception is stated on the last 2 lines of the sample exculpatory language quoted above).

Consequently, you will have to let OSHA have its exception if you want to settle the case with an exculpatory clause. But that's a small price to pay.

OSHA will be pleased to settle a contested citation with the exculpatory language quoted on the preceeding page. That's good news for the employer in most cases. It's **always** better than allowing an OSHA citation to go uncontested.

7. A Typical OSHA Case

If you receive an OSHA citation and you exercise your right to contest it within the 15 working-day time limit, the resulting "case" will unfold at a rather leisurely pace.

Ordinarily, there will be about one year between the time of the inspection and final decision. But the case can be (and usually is) settled at any time during that year.

As an example, a "typical" OSHA case is described on the next 5 pages. The inspection takes place in the Spring of one year and the final decision occurs in the Spring of the following year.

A TYPICAL OSHA CASE

A One Year Voyage From Inspection to Decision -- With Innumerable Opportunities to Settle Along the Way.

1993

April 12 & 13

OSHA conducts inspection at company's plant. The average <u>safety</u> inspection lasts 16 hours. The average <u>health</u> inspection (sampling for employees exposure to air contaminants, noise, etc.) lasts 41 hours. In "state plan" states those numbers are 9 and 27 respectively.

April 13 - May 24

The OSHA inspector works at his office preparing the citations and doing other paperwork. He may contact your equipment suppliers, employees and others for information he needs to justify citation. He may also conduct inspections of other companies. Citations are supposed to issue two days after the inspection ends, but OSHA never does that. Two months later is more like it.

91

May 25	OSHA sends the citations to you by certified mail. Included in the same envelope is (1) an invitation for you to participate in an "informal conference" with OSHA if you dispute the citations or amount of proposed penalty, and (2) a pamphlet advising you of your right to contest the citation and penalty. 

May 25-June 16   If you want to accept OSHA's "informal conference" invitation, it must be done during the 15 "working days" that begin the day after the citation is issued (Saturdays and Sundays are excluded) and the conference must be <u>completed</u> during this period. In this example, the employer decided against the "informal conference." If he had participated in it but was not satisfied with the result, he could still contest the citations in the manner related below.

June 1   Employer mails OSHA a "Notice of Contest" contesting the citations and penalty amounts.

June 9   OSHA mails your Notice of Contest and copies of the citations to the OSH Review Commission, 1825 K Street, N.W., Washington, D.C. The Commission is <u>not</u> affiliated with OSHA in any way, shape or form. At the same time, OSHA turns its files on the case over to the office of the Labor Department's Regional Solicitor. From that point on, all negotiations and other matters will be the responsibility of the Labor Department lawyer assigned to the case by the Regional Solicitor.

June 16   The OSH Review Commission sets up a case file, assigns a docket number to the matter (for example: OSHRC Docket No. 93-1889), and sends computer-generated letter to you and the Labor Department's regional solicitor that states the docket number and encloses a copy of the Commission's procedural rules.

No. 93-1889

June 29   The Labor Department's Regional Solicitor must file with the OSH Review Commission (with a copy to you) a "complaint" which is a legal pleading that sets forth the factual reasons for each OSHA citation (it is usually phrased in "boilerplate" or legal "gobbledygook"). This document is supposed to be filed 20 days after OSHA received your Notice of Contest. However, the Labor Department rarely meets that deadline. You may receive a telephone call from the Labor Department lawyer asking you to agree to an extension of time for the filing of the "Complaint" (until July 29, for example). Ordinarily, requests for extension of time are granted. During the same conversation, the lawyer may also ask if you are interested in settling the case. If you are, the two of you will discuss settlement terms by telephone and, if you reach agreement, the Labor Department's lawyer will prepare and process the necessary paperwork.

July 29   The "Complaint" is filed with the OSH Review Commission by a Labor Department lawyer. A copy is sent to you. Most "state-plan" states omit the "Complaint" and "Answer" requirements.

August 19

You are obligated to file with the OSH Review Commission (with a copy to the Labor Department's lawyer) an "Answer" to the Complaint. The Answer is a legal pleading that is basically a statement denying the allegations made in the "Complaint" (like the "Complaint", it, too, is usually boilerplate and gobbledygook). The Answer is due 20 days after the Complaint is filed (an extension of time will be freely granted if you feel you need one).

September 9

The OSH Review Commission assigns the case to one of its Administration Law Judges (ALJs). The ALJ will later schedule the case for trial, probably within 3 to 6 months from the date of his assignment. He will, however, first notify you and the Labor Department lawyer that the two of you should try to settle the case. If the two of you reach a settlement, a written agreement is filed with the ALJ and the case is over.

July 29-December 31

Both parties will have 5 or 6 months from the time the "Complaint" is filed to conduct pre-trial pleadings, pursuant to the OSH Review Commission's rules. They are set forth in 29 CFR Part 2200 but a copy will be sent to you by the Commission (see June 16 date, above). You can use this procedure to obtain the OSHA inspector's notes that were made during the inspection and all other documentation the agency has collected in support of its case against you. This period of time is most generally used for settlement negotiations between the parties. Approximately 90% of all contested citations are settled prior to trial.

1994

March 7

If the case hasn't been settled, it will go to trial before the ALJ. The place of the trial will be the city where your plant is located (or as close to it as possible). Most trials last one or two days. The ALJ will <u>not</u> decide the case at the close of the trial. He will generally give the parties a couple of months after the trial ends to submit written arguments and briefs to him. He will study them before deciding the case. His decision will be in writing and will explain his reasons in some detail. A period of 6 months between the close of the trial and the ALJ's decision is not unusual (frequently it is longer than that). The party that loses can both petition for review and appeal (but not at the same time). That happens in less than 10% of the cases. At any time <u>during</u> the trial (or even after it is over), the case can be settled if you and the Labor Department lawyer can agree. The ALJ is always happy to have a case settle.

V.   CONSULTATIVE SERVICES

1.   Why They Exist

From day one of OSHA's existence, employers have complained that its very difficult to learn what they are required to do. They don't know what standards apply and, even those employers who know what standards apply often cannot understand the meaning of the requirements.

Congress heard those complaints and did something about it. As a result, government-funded consultative services were established in every state in the Nation (See Part Three of this book where all of them are listed). In fiscal year 1993, Congress appropriated $29.7 million to fund those consultative services.

The **only** reason for existence of the consultative services is to help employers understand OSHA requirements and to explain what must be done to comply. That means you. If you are in doubt about OSHA requirements or you have an OSHA question you want answered, contact your own state's consultative service. Don't be reluctant about it. You are paying for it with your taxes.

2.   How The Consultative Service Operates

Each local consultative service utilizes a cooperative approach to the solution of your occupational safety and health problems. They want to help you. But it's a voluntary activity. **You must request it**.

The service is delivered by State governments using well-trained professional staff. The consultants will answer questions, will help employers identify and correct specific hazards, will provide guidance in establishing or improving an effective safety and health program, and will offer training and education for employers and employees. In short, if you need help or guidance on **any** OSHA matter, the consultative service will provide it.

The consultants make house calls. In fact, the service is given chiefly at your worksite, but limited services (like classroom instruction) may be provided elsewhere.

The safety and health consultation program is **completely separate** from both state and federal OSHA inspection and enforcement.

The service is also **confidential**. Your name and firm and any information about your workplace, plus any unsafe or unhealthful working conditions that the consultant uncovers will *not* be reported to the OSHA inspection staff.* In addition, no citations are issued or penalties proposed as a result of a consultation.

The consultants are reasonable people. If they think something is hazardous but you don't, they will hear you out. Those situations don't occur very often but when they do, they are nearly always worked out to the satisfaction of the employer.

---

*There is one minor caveat. If a consultant observes a condition or practice at your workplace that could reasonably be expected to cause **immediate** death or serious physical harm to employees **and** you steadfastly refuse to take any corrective measures, the consultant is obligated to call that to OSHA's attention. That almost never happens.

3.   Costs and Benefits

The consultants will tell you whether you are in compliance with OSHA requirements. If you aren't, they will tell you what the hazards in your workplace are and the ways to remedy them. You will then be in a much better position.

You can breathe easier once a consultant confirms that you are in compliance. Or you will know exactly what you must do in order to comply with OSHA requirements. That can help you avoid the potential adverse consequences discussed above. It might also reduce your worker comp costs and even win you enhanced respect and appreciation from your employees.

There are additional benefits. The probability of receiving a "willful" OSHA citation in the future is reduced and may well be eliminated. You could also be excluded from "general schedule" OSHA enforcement inspections for one year if you have a complete examination of your workplace, correct all identified hazards, post a notice of their correction, and institute the core elements of an effective safety and health program.

That doesn't mean that you won't get inspected at all. It won't insulate you from inspections that are triggered by employee complaints, fatalities or similar events. But it will nearly always prove that you are dedicated to providing your employees with safe and healthful working conditions -- and that will generally insulate you from willful OSHA violations.

The consultation program provides professional advice and assistance at no cost. There is, however, a potential problem that could arise if you receive safety recommendations -- no matter who or what provides them -- but you do nothing about them. That problem can be avoided if you follow the suggestions under the heading: "Cooperation and Corroboration".

4.   Procedures

All you need to do is call -- or write -- or send a fax. The consultant will make an appointment with you to visit your workplace or will provide you with whatever advice or assistance you request.

The consultative service will do what **you** want done. If you only want guidance on a particular machine or condition, or you only want to know what you should be doing about a certain OSHA standard, the consultant will limit his services accordingly.

If, on the other hand, you want comprehensive consultation services, it will include the following: (1) an appraisal of all mechanical and environmental hazards and physical work practices; (2) an appraisal of your present job safety and health program or the establishment of one; (3) a conference with management on findings; (4) a written report of recommendations and agreements; (5) training, and assistance with implementing recommendations; and (6) a follow-up to see if corrections were properly made or additional help needed.

Consultation starts with your request. If you only want a question answered, one telephone call may be all that's necessary.

When you request onsite consultation, your request will be prioritized according to the nature of your workplace and any backlog of requests that might exist. A consultant assigned to your request will contact you to set up a visit date based upon the priority assigned to your request, your work schedule, and the time needed for the consultant to prepare adequately to serve you.

The consultant may encourage you to include within the scope of your request all working conditions at the worksite and your entire safety and health program. You have the option, however, to limit the consultation visit to a discussion of fewer, more specific problems. If the consultant observes hazards that are outside the scope of the request, he or she will tell you of their presence and discuss them with you.

5.   On-Site Consultation

Upon arrival at your worksite for a scheduled visit, the consultant will briefly review his or her role during the visit and may, if requested, review with you your safety and health program. The consultant will explain the relationship between onsite consultation and OSHA enforcement activity and may discuss your obligation to protect employees in the event that "serious" hazardous conditions exist. Also, he or she will explain that employee participation is encouraged -- but not required -- during the consultation process. It's your option.

After a brief discussion of those matters, there will be a walkthrough of the worksite so that you and the consultant can examine conditions in your workplace. The consultant will identify any specific hazards and provide advice and assistance in establishing or improving your safety and health program and in correcting any hazardous conditions identified.

At your request, assistance may also include education and training programs for you, your supervisory personnel, and other employees.

The consultant should be allowed to talk freely with workers during the walkthrough. That frequently helps him identify and judge the nature and extent of potential hazards.

In a complete review of a company's operation, the consultant will look for mechanical and physical hazards by examining the structural condition of the building, the condition of the floors and stairs, and the exits and fire protection equipment. During the tour of the workplace, he or she will review the layout for adequate space in aisles and between machines, check equipment such as forklifts, and examine storage conditions. Control of electrical hazards and machine guards will also be considered.

101

The consultant will check the controls used to limit worker exposure to environmental hazards such as toxic substances and corrosives, especially air contaminants. He or she will check to see if all necessary technical and personal protective equipment is available and functioning properly, any problems workers may encounter from exposure to noise, vibration, extreme temperatures, lighting, or other environmental factors, and will explain the means and techniques commonly used for the elimination or control of any hazards.

Work practices, including the use, care and maintenance of hand tools and portable power tools, as well as general housekeeping, are frequently of interest to the consultant. He or she may want to talk with you and your workers about items such as job training, supervision, safety and health orientation and procedures, and the maintenance and repair of equipment.

In addition, the consultant will want to know about any on-going safety and health program your firm has developed. If your firm does not have a program or you would like to make improvements, the consultant will, at your request, offer advice and technical assistance on establishing a program or improving it.

Management and worker attitude toward safety and health will be considered in this analysis, as well as current injury and illness data. The consultant may also want to know about how you and your employees communicate about safety and health as well as be aware of any in-plant safety and health inspection programs.

6.   Closing Conference

Following the walkthrough, the consultant will meet with you in a closing conference. This session offers the consultant an opportunity to discuss measures that are already effective and any practices that warrant improvement. During this time, you and the consultant can discuss problems, possible solutions, and time frames for eliminating or controlling any serious hazards that were identified during the walkthrough.

Consultants offer general approaches and options as well as technical assistance on the correction of hazards. As necessary, consultants will recommend other sources for specialized technical help.

The consultant may also offer suggestions for establishing, modifying, or adding to the company's safety and health program to make such programs more effective. Such suggestions could include worker training, changing work practices, methods for holding supervisors and employees accountable for safety and health, and various methods of promoting safety and health.

After the closing conference, the consultant will send you a written report explaining the findings and confirming any correction periods that the two of you may have agreed upon. The report may also include suggested means or approaches for eliminating or controlling hazards as well as recommendations for making your safety and health program effective.

You are free to contact the consultants for additional assistance at any time.

7. The Certificate Option

Employers who receive a comprehensive, on-site consultation visit, correct all identified hazards, and institute the core elements of an effective safety and health program can be awarded a certificate of recognition from OSHA signifying a 1-year exemption from general schedule enforcement inspections. That is an option you can choose or refuse. It's strictly up to you.

Employers who elect to pursue this exemption must post a notice of participation as well as notice of correction of all of the hazards that the consultant found.

Once the certificate is awarded, participants must agree that they will call the consultant for further assistance if new processes that could pose new hazards are introduced into the workplace. They must also agree to complete the remaining elements of an effective safety and health program within a reasonable time (if the consultant identified any).

At present, the exemption provision applies only to fixed worksites. The inclusion of construction and other mobile worksite is under review. This exemption provision applies only in states under Federal OSHA enforcement. Similar provisions may be adopted by the state plan states.

If that option interests you, get the full details from the consultant.

8. Cooperation and Corroboration

There are many ways that the consultative service can help you with OSHA problems. One of them has already been mentioned: Avoidance of "willful" violation citations in future OSHA inspections. A "willful" violation occurs when the employer demonstrates **careless indifference** to employee safety. The act of seeking out a consultative service in order to improve worker safety is the very antithesis of "careless indifference."

A Vermont OSHA leaflet states that: "Contacting [the Vermont consultative service] for free assistance is clear evidence to [Vermont OSHA] that you are interested in complying with the law voluntarily." That's also true everywhere else.

If you know exactly what help or assistance you want, you can make that clear in your initial contact with the consultative service and the consultant's response will be limited accordingly. If, on the other hand, you want a complete, on-site audit of your workplace, that can be arranged in the manner stated above.

In either case, however, you should fully cooperate with the consultant and you should make some kind of a record or notation of the entire process -- from your initial request to what you did (or did not do) in response to the consultant's recommendations.

That will be relatively easy in most cases. You will have a copy of your written request for assistance. The consultant will submit a written report to you and you will take steps to implement his recommendations by ordering supplies or equipment, or by making repairs or changes. You will probably have your own work orders or similar documents that will be there to show what was done.

There will be other situations, however, where you disagree with a consultant's recommendations or -- for one reason or another -- fail to carry them out. That could cause you future problems if you leave things hanging. Those problems won't come from the consultative service. The consultant can only recommend. He can't force you to implement his recommendations -- and he probably wouldn't want to force anything on you. The more likely source of potential problems for you is a future accident or similar event. Here's an example:

Your consultant identifies a potentially hazardous condition or practice and recommends a change. You consider it and decide it to be impractical or not feasible. At a later date, an accident occurs that involves the same condition or practice. An investigation or a lawsuit is initiated in an effort to prove your responsibility for the tragedy. The fact of the recommendation -- and your failure to implement it -- is uncovered. It is then alleged that the tragedy could have been avoided if you had implemented the recommendation. You are then portrayed as a person who is carelessly indifferent to safety or one who willfully refused to do the right thing because you placed your own desire for profit above the safety and health of your employees.

That scenario is not a figment of someone's imagination. It has occurred over and over again throughout history. It has also been the basis for numerous OSHA "willful" violations and multi-million dollar OSHA fines.

The recommendation for change doesn't have to come from an OSHA consultant. It could come from within your own business, or from your insurance carrier, a salesman, or from virtually anywhere. The adverse implications that can result from the failure to follow a safety recommendation, however, is the same.

The lesson to be learned from that phenomena is simple: **Always Respond To Safety Recommendations.**

The "response" doesn't have to be the implementation of the recommendation. If there are good and valid reasons why the recommendation is impractical or should not be put into effect -- and that is not an unusual situation -- those reasons should be **documented**. Put them in writing by either a letter to the person who made the recommendation or by a memorandum of some kind that you keep in the same file as the recommendation.

You must do something that can later be used to show that you didn't simply carelessly disregard a safety recommendation.

If, on the other hand, you accept a consultant's recommendation and put it into effect, that fact should also be documented. Keep a record of what you did, with copies of invoices or work orders. That documentation can be used in the event someone later raises questions about the matter.

There will be still other occasions when you orally ask for (and receive) advice from a consultant -- or even from OSHA itself. You should also make some kind of record or notation of that -- for the same reasons discussed above. Make a notation of the calendar date of the conversation, the name of all persons who participated in it, who said what, what you did (or did not do) as a result and, if you did not accept the advice, the reason why.

You can keep notations of that kind confidential if you so desire. But **do** make a record for your own use. It could prove to be invaluable in the future if you choose to use it. But, if you don't make a notation, and a future dispute arises over who said what to whom and when it happened, you may find that memories have faded or people recall the same event differently.

Never leave any loose ends when dealing with a consultant who recommends changes in the interests of safety.   Respond to **every** suggestion and recommendation.

If you disagree with the consultant, and through use of logic are able to persuade him to alter the original recommendation, **make sure the consultant makes some kind of written acknowledgement of that change.** That should also be done with all other safety recommendations irrespective of who made them.

The matters discussed above apply equally to private consultants you may employ, safety advice you might seek from any source, inspections or consultative visits that might be made by your insurance carrier or a public agency, as well as recommendations that come from sales representatives or persons within your own business, or from an employee safety committee, or a labor union that represents your employees.

There's a simple rule to follow: **Always Respond To Every Safety Recommendation and Retain Documentation of Your Response.**

9. **Summary**

The consultation program provides several benefits for you as an employer.

Onsite consultants **will**:

- Help you recognize hazards in your workplace.

- Suggest approaches or options for solving a safety or health problem.

- Identify sources of help available to you if you need further assistance.

- Provide you with a written report that summarizes their findings.

- Assist you in developing or maintaining an effective safety and health program.

- Offer training and education for you and your employees at your workplace, and in some cases away from the site.

- Under specified circumstances, recommend you for recognition by OSHA and a 1-year exclusion from general schedule enforcement inspections.

Onsite consultants will **not**:

- Issue citations or propose penalties for violations of Federal or State OSHA standards.

- Routinely report possible violations to OSHA enforcement staff.

- Guarantee that any workplace will "pass" a Federal or State OSHA inspection.

You should always:

- Cooperate fully with the consultant.

- Respond to every recommendation and retain documentation of your response.

The consultation program is there to serve you. It won't cost you anything. It might protect you from "willful" OSHA citations and the huge fines that go with them. It could be the answer to your problems and provide you with the peace of mind that comes from knowing that you are doing things correctly. Give it a try.

PART TWO

OSHA COMPLIANCE

HOW TO UNDERSTAND THE REQUIREMENTS

I.   INTRODUCTION TO THE OSHA REQUIREMENTS

1.   Overview

Your OSHA obligations as an employer fall into three general categories:

- A general duty to maintain a workplace free from recognized hazards that are likely to cause death or serious physical harm to your employees. That is a part of the OSH Act known as the "general duty clause". Its wording is rather ambiguous and indefinite. It is not used very often as the basis for citation and it is not supposed to be. Congress directed OSHA to adopt specific **standards** to control workplace hazards.

- Observe all applicable occupational safety and health standards ("OSHA standards") promulgated by the Secretary of Labor. There are thousands of OSHA standards. What they cover and who must comply with them is explained in the next section. Over 95% of all OSHA citations to date have alleged violations of OSHA standards. Understanding and observing them is, therefore, the most important of the three employer responsibilities listed here.

- Virtually every employer is obligated to keep **records** of their employees' "recordable" injuries and illnesses, report work-related employee fatalities and multiple hospitalizations to OSHA, and display an OSHA-supplied Poster that provides general information on the OSH Act. Those rules are summarized later in this book. In addition, many employers must observe hundreds of other OSHA recordkeeping and reporting requirements that are included in the "OSHA standards." An explanation of what they cover and who must comply is included in the section on OSHA standards that follows.

The OSHA obligations of employers located in the 21 "State Plan" states are, for the most part, identical to those stated above. There are, however, some differences in some states.

The OSH Act requires that the OSHA standards in state plan states be "at least as effective" as those adopted by OSHA itself. With few exceptions, the standards in those states are **identical** to OSHAs. They even use the same "CFR" designations.

Some state plan states, however, have gone beyond OSHA in the adoption of regulations -- they have adopted all OSHA standards and added some of their own. For example, there is **no** OSHA requirement that employers implement comprehensive workplace safety and health programs. But there are seven states that impose such a requirement: Alaska, California, Hawaii, Minnesota, Oregon, Texas and Washington.

You can find out if your state has safety and health rules that differ from the OSHA standards by contacting the consultative service for your state listed in Part Three of this book.

2. The OSHA Standards -- In General

OSHA standards have the same status and effect as regulations adopted under other federal laws -- the Internal Revenue Code, for example. You must comply with them, or you can be penalized by citations and fines.

Shortly after the OSH Act went into effect in 1971, the Secretary of Labor, under the authority delegated by Congress, adopted thousands of occupational safety and health standards. In subsequent years, additional standards have been added. Some of those standards have since been revised while others have been revoked.

The Secretary's authority to adopt OSHA standards is a continuing one. Thus, new standards can be adopted in the future.

Job safety and health standards generally consist of rules for avoidance of hazards that have been proven by research and experience to be harmful to personal safety and health. The standards supposedly constitute an extensive compilation of wisdom. They sometimes apply to all employers, as do fire protection standards, for example. A great many standards, however, apply only to workers while engaged in specific types of work-such as driving a truck or handling compressed gases.

Two of the many thousands of occupational safety and health standards are listed below in order to demonstrate the form of such standards.

Example No. 1

"Aisles and passageways shall be kept clear and in good repair, with no obstruction across or in aisles that could create a hazard."

Example No. 2

"Employees working in areas where there is a possible danger of head injury from impact, or from falling or flying objects, or from electrical shock and burns, shall be protected by protective helmets."

It is the obligation of all employers and employees to familiarize themselves with the standards that apply to them and to observe the standards at all times.

Once an OSHA standard has been adopted, it is published in the Code of Federal Regulations (CFR). The CFR is divided into 50 "titles" that cover **all** regulations adopted by **all** federal agencies. Each "Title" is designated by a number, beginning with "1" and ending with "50". The OSHA standards are part of "Title 29" the section of the CFR assigned to "Labor" regulations.

Title 29 is further subdivided into various "Parts" covering specific regulatory areas. All of the OSHA standards that apply to private employers are included in nine parts as follows:

- Part 1903   Inspections, citations and proposed penalties

- Part 1904   Recording and reporting occupational injuries and illnesses

- Part 1910   Occupational safety and health standards for *General Industry*

- Part 1915   Occupational safety and health standards for *Shipyard Employment*

- Part 1917   Occupational safety and health standards for *Marine Terminals*

- Part 1918   Occupational safety and health standards for *Longshoring*

- Part 1919   *Gear certification* rules (maritime industry)

- Part 1926   Occupational safety and health standards for *Construction*

- Part 1928   Occupational safety and health standards for *Agriculture*

The explanation of those standards in this book is grouped into 5 categories:

1. Recordkeeping/Reporting (Parts 1903 and 1904)[1]

2. General Industry (Part 1910)

3. Maritime (Parts 1915, 1917, 1918 and 1919)

4. Construction (Part 1926)

5. Agriculture (Part 1928)

The Maritime, Construction and Agriculture standards apply to companies engaged in those industries. The "General Industry" standards apply to all other businesses and industries. The following illustration is offered as an explanation:

---

[1]As used in this book, the term "Recordkeeping/Reporting" means a regulation that imposes some kind of paperwork requirement upon employers. Parts 1903 and 1904 only contain some of those requirements. Many additional ones are contained within the requirements of particular OSHA standards. For example, the OSHA noise standard includes requirements that **records** be made of noise exposure measurements and hearing test results.

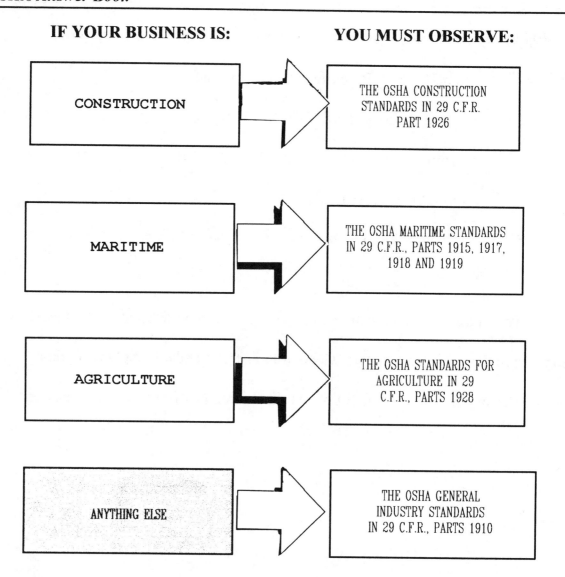

IF YOUR BUSINESS IS: **YOU MUST OBSERVE:**

CONSTRUCTION → THE OSHA CONSTRUCTION STANDARDS IN 29 C.F.R. PART 1926

MARITIME → THE OSHA MARITIME STANDARDS IN 29 C.F.R., PARTS 1915, 1917, 1918 AND 1919

AGRICULTURE → THE OSHA STANDARDS FOR AGRICULTURE IN 29 C.F.R., PARTS 1928

ANYTHING ELSE → THE OSHA GENERAL INDUSTRY STANDARDS IN 29 C.F.R., PARTS 1910

Most businesses, of course, will fall into the "anything else" category listed above. The Part 1910 "General Industry" standards apply to far more employers than do the construction, maritime or agriculture standards. There are even some limited situations in which they apply to employers engaged in construction, maritime and agriculture. The particular OSHA standards that you should become familiar with depends upon the business you are in.

But there is one more obligation that must not be overlooked. No matter what business you are in, you must observe the:

OSHA RECORDKEEPING AND REPORTING REGULATIONS IN 29 C.F.R. PARTS 1903

There are fifteen (15) recordkeeping and reporting regulations that apply to all employers. Each of them is listed in this book. All employers (with some rather limited exceptions, discussed later) must observe those recordkeeping/reporting regulations.

In addition, those employers who are engaged in construction must comply with the Construction standards, employers in the maritime industries must observe the Maritime standards. Those engaged in agriculture must comply with the Agriculture standards, and those engaged in any other kind of business must observe the General Industry standards.

It should also be noted that all of the recordkeeping/reporting regulations are not included in Parts 1903 and 1904. There are nearly 400 other recordkeeping and reporting (or employer paperwork) requirements that are included in various OSHA **standards**. They require the employer to make, obtain and keep certain records or documents, or to file reports, post notices, give warnings, provide information, or do similar things. They apply if the OSHA standard (that they are a part of) applies.

For example, if you have a respirator on your premises, the OSHA respiratory protection standard applies to you. It contains employer paperwork requirements. If you use electricity or machines in your business, the OSHA electrical and/or machinery standards apply to you. They also include recordkeeping requirements.

This book is designed to help you find out what the requirements are. But you will also need a copy of the OSHA standards.

3.　You Must Have A Copy Of The Standards

The first thing you must do, therefore, is to get a copy of the OSHA standards that apply to you. You'll find them in the Code of Federal Regulations unless your business is located in a "state plan" state.

Turn to Part Three of this book for the page or pages for your state. If it says on the first line "Enforcement and administration by OSHA", you must get a copy of the appropriate volume or volumes of the Code of Federal Regulations. If you don't already have a copy, telephone a nearby OSHA office (addresses and telephone numbers are listed in Part Three of this book). They will either give you a copy free of charge or they will tell you how to get the volume (or volumes) you need.

If your business is in a "state plan" state (turn to your own state's listing in Part Three of this book to determine that), telephone one of the state OSHA enforcement or consultation offices listed for your state. They will either give you a free copy of the applicable state OSHA requirements or will tell you how to get them.

BE SURE TO GET A COPY OF THE
STANDARDS. THERE MUST BE A
COPY PHYSICALLY PRESENT AT
EVERY LOCATION WHERE PEOPLE
WORK.

It's very easy to get a copy. Pick up your telephone. In most cases, that's all you have to do.

If you do not have a copy of the standards, you will almost certainly be in violation of them. There are a number of OSHA standards, for example, that require the employer to make the text of the standard available for their employees. You are in **violation** of those standards if you do not have a copy available at each of your workplaces. See, for example 29 C.F.R. §1910.20(g)(2) which is listed infra in the third section of this book.

Also, without a copy of the standards, you won't be able to fully comprehend the matters discussed in this book. So, do it **NOW**.

It is not necessary that you read every standard and, even if you were so inclined, you almost certainly would not fully understand them. But, you need to have them at hand. One purpose of this book is to point you in the direction of OSHA requirements that apply to you and to tell you who to call for help in understanding or observing those requirements.

You need a copy of the standards in order to know that there are OSHA standards that cover particular aspects of your business. Knowing that is the first step. Understanding those standards comes later. But you cannot take the first step unless you have your own copy of the standards. So don't put it off. Get your copy (or copies) at once.

The OSHA standards are bound in paperback books. If you want **all** OSHA standards and regulations, you will have to get five separate volumes, but if you only want the construction standards or the maritime standards, you'll only need one book (general industry standards cover 2 books).

They can also be purchased from the U.S. Government Printing Office directly or from one of its bookstores. There are also some private publishers who sell copies of the OSHA standards in book form.

The OSHA standards and regulations are also available on computer disks from various merchants and from the U.S. Government Printing Office.

4. How To Read An OSHA Standard

You don't have to read all the standards -- but you will want to read a particular standard at some point in time. The following discussion can help you do that.

Many OSHA standards are lengthy and complex. The difficulty of reading them is complicated by the use of small type to print them and the failure to separate each subsection by some readily identifiable method or a reader-friendly format.

All OSHA standards, however, are arranged in the same format or numbering system. Once that format is understood, locating particular subsections of a standard will be easier. The next few paragraphs will help you to understand that.

Publication of the Code of Federal Regulations (CFR) is the responsibility of the Director of the Office of the Federal Register. He decides where in the CFR each federal agencies' regulations will be published. He has assigned "Title 29" to OSHA. Thus each OSHA regulation is preceded by "29 C.F.R."

Individual **parts** of Title 29 are designated by a four digit number (for example, 1901, 1903, 1910, 1926). That "part" designation is the first four numbers of all OSHA standards. A period follows the first four digits. Following that period, each standard is listed in numerical order beginning with number one. Thus 29 C.F.R. 1910.1 (or, as designated in this book, 29 C.F.R. §1910.1) refers to Title 29, Part 1910, Section 1.

The number following the period is the designation given to a particular OSHA standard. In many cases it will be followed by a name. For example:

§1910.268 Telecommunications

There then will follow various subsections of the standard. They are designated by letters and numbers **all of which are in parenthesis**.

The first subsection is "(a)". For example §1910.268(a). Other major subsections will be similarly designated in alphabetical order. For example, §1910.268(b), §1910.268(c), etc. However, each of them often has its own subsections. And those subsections, in turn, have their own subsections.

The **subsections** are arranged and designated as follows:

First Group: alphabetically by small letters as described above

Second Group: numerically. For example, (1), (2), (3), etc.

Third Group: numerically by Roman numerals. For example (i), (ii), (iii), etc.

Fourth Group: alphabetically by capital letters. For example, (A), (B), (C), etc.

For subsequent groups (if necessary), the order listed above is repeated but the letter or number is in italics. For example *(a)*, *(1)*, etc.

An example to demonstrate the foregoing discussion follows.

Listed Below Are The Steps to be Taken to Locate:

29 C.F.R. §1910.268(j)(4)(iv)(F)

1.   Get a copy of Title 29, C.F.R., Parts 1900 to 1910

2.   Locate §1910.268. Telecommunications

3.   Begin with "(a)", then look to (b), (c), (d), etc. until you find "(j)"

4.   Find "(1)" under subsection "(j)". Then look for "(2)", "(3)" and "(4)"

5.   Once you have located "(4)", look for Roman numeral "(i)", then for "(ii)", (iii)", "(iv)"

6.   Once you have located "(iv)", look for capital letter "(A)", then find "(B)", "(C)", "(D)" etc., until you have located "(F)"

The format and sequence for all OSHA standards is the same. If those same steps are followed, you should be able to locate the particular subsection of the standard being sought.

II.   OSHA RECORDKEEPING AND REPORTING REGULATIONS

1.   Employee Injuries and Illnesses

There are fifteen recordkeeping and reporting requirements that apply to nearly all employers, the most common of which is the recording of employee injuries and illnesses as they occur.

They must be recorded on OSHA-supplied forms (or their equivalent) on a calendar year basis. Entries must be made on the forms at the time employee injuries and illnesses occur. The completed forms are not sent to OSHA or anyone else. But they must be kept for five years and be available for inspection.

Forms

Those injury/illness recordkeeping forms are the **OSHA Form 200** and the **OSHA Form 101**. If you do not have the forms, telephone your nearest OSHA or state-OSHA office. They will furnish you with the forms and a pamphlet that explains your injury/illness recordkeeping obligation.

The injury/illness recordkeeping requirements apply to all employers of 11 or more employees, except for some employers engaged in certain retail trades and in certain service industries.[2]

> The exemption for small employers (10 or fewer employees) applies only to injury/illness recordkeeping. Those exempt employers must, like all other employers, comply with all applicable OSHA standards (including their recordkeeping provisions), display the OSHA poster, and report fatal accidents and multiple hospitalizations to OSHA.

If an on-the-job accident occurs that results in the death of an employee or in the hospitalization of three or more employees, that employer, regardless of the number of employees he has, must report the accident to the nearest OSHA office within 8 hours.

In state-plan states, employers report such accidents to the state agency responsible for safety and health programs. Check the state-by-state listings in Part Three of this book for the nearest office.

---

[2] Employers with ten or fewer full or part-time employees are generally exempted from the injury/illness recordkeeping requirement, as are employers engaged in retail trade, finance, insurance and real estate. See 29 C.F.R. §§1904.15(a) and 16. Establishments classified in Standard Industrial Classification Codes (SIC) 52-89 are also exempted, with the exception of those in SIC 52 (building materials and garden supplies), 53 and 54 (general merchandise and food stores), 70 (hotels and other lodging establishments), 75 and 76 (repair services), 79 (amusement and recreation services) and 80 (health services). However, those exempted employers are obligated to report to OSHA any incident resulting in the death of an employee or the hospitalization of five or more employees. 29 C.F.R. §1904.16. Those employers also may be obligated to keep records if they are selected to serve as the subject of a Bureau of Labor Statistics (BLS) survey. 29 C.F.R. §1904.21. If any business or industry is so selected, it will be personally advised by BLS.

2.   The Particular Requirements

A complete listing of the 15 generally-applicable recordkeeping and reporting

requirements follows.

> ### EMPLOYERS ARE REQUIRED TO OBSERVE THE FOLLOWING
> ### OSHA RECORDKEEPING AND REPORTING REQUIREMENTS

---

29 C.F.R. §1903.2(a)	The OSHA poster advising employees of the OSH Act's provisions must be posted in each establishment.
29 C.F.R. §1903.16(a)	Whenever OSHA citations are received by the employer, they must be posted at or near each location of alleged violation.
29 C.F.R. §1904.2	Employers must keep a log and summary of all recordable occupational injuries and illnesses sustained by their employees (OSHA 200 Form).
29 C.F.R. §1904.4	Employers must also keep supplementary records containing additional details on each recordable injury and illness (OSHA 101 Form).
29 C.F.R. §1904.5	An annual summary of all recordable injuries and illnesses must be posted from February 1 to March 1 of each year.
29 C.F.R. §1904.6	The employee injury/illness records must be maintained for 5 years.
29 C.F.R. §1904.7	Each employer must provide each employee with access to his own injury/illness records.
29 C.F.R. §1904.8(a)	Within 8 hours after the death of any employee from a work-related incident or the in-patient hospitalization of three or more employees as a result of a work-related incident, the employer of any employees so affected must report the fatality/multiple hospitalization by telephone or in person to the Area Office of the Occupational Safety and Health Administration (OSHA), U. S. Department of Labor, that is nearest to the site of the incident, or by using the OSHA toll-free central telephone number.

29 C.F.R. §1904.8(b)	An employer must report each such fatality or hospitalization of three or more employees which occurs within thirty (30) days of an incident.
29 C.F.R. §1904.8(c)	Exception: If the employer does not learn of a reportable incident at the time it occurs and the incident would otherwise be reportable under paragraphs (a) and (b) of this section, the employer shall make the report within 8 hours of the time the incident is reported to any agent or employee of the employer.
29 C.F.R. §1904.8(d)	Each report should have the following information:Establishment name, location of incident, number of fatalities or hospitalized employees, contact person, phone number, and a brief description of the incident.
29 C.F.R. §1904.11	Employer is responsible for maintaining Records/Reports of prior owner on OSHA matters when a change of ownership occurs.
29 C.F.R. §1910.20(d)	If the employer keeps such records, employee **medical** records must be kept for the duration of employment plus 30 years and employee **exposure** records must be kept for 30 years unless otherwise provided in some other OSHA standard.
29 C.F.R. §1910.20(e)	If the employer keeps employee exposure and medical records, he must provide his employees with access to their records upon request.
29 C.F.R. §1910.20(g)(1)	If there are employees who are exposed to toxic substances or harmful physical agents, they must be notified when hired and at least annually thereafter of their right of access to their medical and exposure records.
29 C.F.R. §1910.20(g)(2)	Employers must (1) keep a copy of the OSHA Records Access Standard, 29 C.F.R. §1910.20, readily available for employees and (2) distribute informational materials to them whenever OSHA provides it.

[3]

---

[3]There are nearly 400 additional recordkeeping and reporting requirements included in various OSHA standards. The applicability of each of them varies. For example, if there is present on your premises any paint, or any other product that OSHA regards as a "hazardous chemical", you must have certain records and documents that are specified in the OSHA Hazard Communication Standard. A complete listing and description of **all** OSHA recordkeeping and reporting requirements is contained in the book, THE OSHA 500, a publication of Moran Associates. Call 1-800-597-2040 for more information.

III. OSHA GENERAL INDUSTRY STANDARDS

1. Overview

The OSHA "general industry" standards are printed in 29 C.F.R. Part 1910. There are a lot of them. Two separate volumes of the Code of Federal Regulations are needed to print them all.

You don't have to read them all. Read over the titles for each Subpart. All of the Subpart titles are listed below. Some will obviously not be applicable to your business. Forget about them. Concentrate on the ones that might be applicable. Then read the discussion of those subparts in the next few pages of this book.

Part 1910 is divided into 21 subparts each of which is designated by a capital letter and a title. Two of them (A and B) do not contain any applicable substantive requirement. The remaining 19 are listed below:

- Subpart C -   General Safety and Health Provisions[4]
- Subpart D -   Walking-Working Surfaces
- Subpart E -   Means of Egress
- Subpart F -   Powered Platforms, Manlifts, and Vehicle-Mounted Work Platforms
- Subpart G -   Occupational Health and Environmental Control
- Subpart H -   Hazardous Material
- Subpart I -   Personal Protective Equipment
- Subpart J -   General Environmental Controls
- Subpart K -   Medical and First Aid

---

[4]The **only** OSHA standard that was in Subpart C was 29 C.F.R. §1910.20, Access to Employee Exposure and Medical Records. That standard has been moved to Subpart 2, Toxic and Hazardous Substances. Consequently, there will be no further discussion of Subpart C in this section

- Subpart L -   Fire Protection
- Subpart M -   Compressed Gas and Compressed Air Equipment
- Subpart N -   Materials Handling and Storage
- Subpart O -   Machinery and Machine Guarding
- Subpart P -   Hand and Portable Powered Tools and Other Handheld Equipment
- Subpart Q -   Welding, Cutting and Brazing
- Subpart R -   Special Industries
- Subpart S -   Electrical
- Subpart T -   Commercial Diving Operations
- Subpart Z -   Toxic and Hazardous Substances

Each Subpart is further broken down into sections. On the following pages, the subparts listed above will be discussed separately (beginning with subpart "D" and continuing in alphabetical order).

Readers should note that there is no Subpart U, V, W, X or Y in the OSHA general industry standards.

2.   Subpart D. Walking-Working Surfaces

The various OSHA standards included in "Subpart D - Walking-Working Surfaces" are shown in the following chart.

SUBPART D - WALKING-WORKING SURFACES	
1910.21	Definitions
1910.22	General requirements
1910.23	Guarding floor and wall openings and holes
1910.24	Fixed industrial stairs
1910.25	Portable wood ladders
1910.26	Portable metal ladders
1910.27	Fixed ladders
1910.28	Safety requirements for scaffolding
1910.29	Manually propelled mobile ladder stands and scaffolds (towers)
1910.30	Other working surfaces
1910.31	Sources of standards
1910.32	Standards organizations

The last 2 listed sections (§1910.31 and §1910.32) are not OSHA "standards". They simply identify the sources from which the standards in the other sections (§1910.21 through §1910.30) were obtained. You'll find similar sections at the tail end of most other Subparts. Those sections are useful for historical and research purposes but otherwise there is no need to be concerned with them.

For the most part, you can tell by looking at the title of each section whether it contains any OSHA standards that you ought to be concerned about. If you have ladders or scaffolds, for example, you will want to read those sections to see what is required and whether the requirements apply to you. You'll need a copy of the Part 1910 standards.

If, after reading the pertinent standards, you are unsure of their meaning, turn to the listing for your state in part three of this book. Locate the appropriate telephone number under "CONSULTATION", pick up the telephone and discuss the matter.

Two of the sections of "Subpart D" have titles that are not very informative: §1910.22 "General Requirements" and §1910.30" "Other Working Surfaces". You can find out the particulars by consulting your copy of the Part 1910 standards.

It will take only a few seconds to determine if there is anything in those two sections that concerns you. Each of those standards cover only half a page in the Code of Federal Regulations (CFR).

3.   **Subpart E. Means of Egress**

SUBPART E - MEANS OF EGRESS	
1910.35	Definitions
1910.36	General requirements
1910.37	Means of egress, general
1910.38	Employee emergency plans and fire prevention plans
1910.39	Sources of standards
1910.40	Standards organizations

If your employees work in a building, this Subpart applies to you. Get a copy and read it. It's only 5 or 6 pages long. Its purpose is to ensure that when people need to have a safe and efficient means of leaving a building under emergency circumstances, the means will be there -- and they will have minimal problems finding it and using it.

Egress: As defined by Webster: "A place or means of going out."

Subpart E contains definitions of terms related to the topic (§1910.35), general requirements which are fundamental to safe and efficient egress from facilities (§1910.36), and detailed requirements to ensure that the general provisions are properly implemented (§§1910.37 and 1910.38).

In addition, there are brief sections on the requirements for exit markings and signs. The latter are covered more completely in Subpart L, entitled "Fire Protection".

While escaping from fires is certainly a primary reason for emergency egress from a building, it is not the only reason. Additional hazards which must be considered include:

- Explosion
- Earthquake
- Smoke (without fire)
- Toxic vapors
- Bomb threat
- Storms (tornado, hurricane, etc.)
- Flash floods
- Nuclear radiation exposure
- Actions or threatened actions of terrorist groups, and similar persons
- Other reasons

Each of those hazards to the occupants of a building can occur singly or in combination with others. Depending on the hazards, the people involved, the characteristics of the building, and the quality of the means of egress provided, each hazard can be compounded by:

- Panic and confusion
- Poor visibility
- Lack of information; misinformation

These compounding factors frequently cause more injuries and fatalities than the hazard itself. Providing the proper means of egress can enable persons to successfully escape from the primary hazard.

Subpart E was "copied" or "borrowed" from a voluntary standard that's been around a long time.

The entire Subpart E was adopted from NFPA (National Fire Protection Association) Code 101-1970, *Life Safety Code*. That Code was prepared, and published by the National Fire Protection Association, Quincy, Massachusetts in 1970.

Your compliance with most requirements of Subpart E has probably been checked by local authorities because the NFPA Life Safety Code is also the basis for most local fire codes.

Subpart E has been written for general applicability. Keep in mind that your concern is its application for the protection of **employees**, not the preservation of facilities. *All OSHA* standards are limited in scope to the protection of employees from workplace hazards.

With your copy of the Part 1910 standards in hand, take a couple of minutes to peruse the first four sections of Subpart E. You may also want to read the 2-page Appendix that follows §1910.40. It explains the requirements for Emergency Action Plans and for fire prevention. If you have any doubts or questions about Subpart E or your compliance status, telephone the state consultation service for your state. You'll find it listed in part three of this book.

4. Subpart F. Powered Platforms

SUBPART F - POWERED PLATFORMS, MANLIFTS, AND VEHICLE-MOUNTED WORK PLATFORMS	
1910.66	Powered platforms for building maintenance
1910.67	Vehicle-mounted elevating and rotating work platforms
1910.68	Manlifts
1910.69	Sources of standards
1910.70	Standards organizations

The Subpart F standards are of rather limited application. They apply only to employers who use the particular kinds of equipment listed in its title during building maintenance operations. If you do, you should read over those standards that cover your equipment, as well as the explanatory appendices that follow §1910.66. It includes three drawings to help you understand the standard.

If you are not involved with the listed equipment, you don't have to concern yourself with Subpart F.

5.   Subpart G. Health and Environmental Controls

SUBPART G - OCCUPATIONAL HEALTH AND ENVIRONMENTAL CONTROL	
1910.94	Ventilation
1910.95	Occupational noise exposure
1910.97	Nonionizing radiation
1910.98	Effective dates
1910.99	Sources of standards
1910.10	Standards organizations

There are only 4 OSHA standards in Subpart G. The one of widest application is the noise standard, §1910.95. It establishes a 90 decibel limit on employee noise exposure when averaged over an 8-hour workday, and requires various protective measures at and above the 85 decibel level. If you have **any** employees working where the noise is at or above those limits, you should read §1910.95. It contains numerous, detailed requirements (including paperwork requirements). If you have questions about the standard, call your state consultation service.

There are 3 other standards in Subpart G. Two of them are limited in application to places where there exists ionizing radiation (such as x-rays), §1910.96 has been moved to Subpart 2, Toxic and Hazardous Substances, and redesigned as §1910.1096, or nonionizing radiation (like laser equipment), §1910.97.

The standard entitled "ventilation", §1910.94, is not as broad in scope as its title suggests. It applies to 4 types of operations:

- Abrasive Blasting Operations
- Grinding, Polishing and Buffing
- Spray Finishing
- Open-Surface Tanks

If your employees are engaged in any of those operations, you should familiarize yourself with the applicable provisions of §1910.94, as follows:

§1910.94(a) for abrasive blasting.

§1910.94(b) for grinding, polishing and buffing.

§1910.94(c) for spray finishing.

§1910.94(d) for open-surface tanks.

There are additional spray finishing requirements included in §1910.107 of Subpart H **if** the spray finishing operation involves the use of flammable or combustible materials. An explanation of that standard is included in the following Subpart H discussion.

6.   Subpart H. Hazardous Materials

SUBPART H. HAZARDOUS MATERIALS	
1910.101	Compressed gases (general requirements)
1910.102	Acetylene
1910.103	Hydrogen
1910.104	Oxygen
1910.105	Nitrous oxide
1910.106	Flammable and combustible liquids
1910.107	Spray finishing using flammable and combustible materials
1910.108	Dip tanks containing flammable or combustible liquids
1910.109	Explosives and blasting agents
1910.110	Storage and handling of liquified petroleum gases
1910.111	Storage and handling of anhydrous ammonia
1910.114	Effective dates
1910.115	Sources of standards
1910.116	Standards organizations
1910.119	Process Safety Management of Highly Hazardous Chemicals
1910.120	Hazardous waste operations and emergency response

Subpart H includes 9 standards that are limited in application and 4 that apply more broadly. If you use any of the following substances or processes in your business, you should familiarize yourself with the relevant OSHA standard listed on the same line.

- Compressed gases   §1910.101
- Acetylene   §1910.102
- Hydrogen   §1910.103
- Oxygen   §1910.104
- Nitrous Oxide   §1910.105
- Dip Tanks   §1910.108
- Explosives and Blasting Agents*   §1910.109
- Liquified Petroleum Gases   §1910.110
- Anhydrous Ammonia   §1910.111

The application of the other 4 standards in Subpart H is much broader in scope. They cover many different substances.

§1910.106 is essentially the same as the National Fire Protection Association's publication: NFPA 30, *Flammable and Combustible Liquids Code*. The standard applies to the handling, storage, and use of flammable and combustible liquids with a flash point below 200°F.[5]

There are two primary hazards associated with those flammable and combustible liquids: explosion and fire. In order to prevent those hazards, the standard addresses the primary concerns of: design and construction, ventilation, ignition sources, and storage.

---

[5]As the result of an amendment adopted in 1992, manufacturers of explosives and pyrotechnics must also observe the requirements of the Process Safety Management standard in §1910.119. See 57 Federal Register 6403, February 24, 1992.

If there are liquids of that kind in your business, you should closely examine and observe the many details included in §1910.106. If you have difficulty understanding the standard, contact your state consultative service for assistance.

§1910.107 applies to places of business where spray finishing is performed (those operations are also regulated by §1910.94(c), discussed above under Subpart G).

§1910.107 regulates flammable and combustible finishing materials when applied as a spray by compressed air, "airless" or "hydraulic atomization," steam electrostatic methods, or by any other means in continuous or intermittent processes. It also covers the application of combustible powders by powder spray guns, electrostatic powder spray guns, fluidized beds, or electrostatic fluidized beds.

The standard does not apply, however, to outdoor spray application of buildings, tanks, or other similar structures, nor to small portable spraying apparatus that is not used repeatedly in the same location. In those situations, there would be lesser chance of combustible residue buildup and greater chance of atmospheric dilution of flammable vapors.

There are two recently- adopted standards included in Subpart H: §1910.119 (process safety management) and §1910.120 (hazardous waste and emergency response). Those 2 standards apply as follows:

Hazardous Waste and Emergency Response. The title of the OSHA standard regulating hazardous waste and emergency response operations, §1910.120, is often misunderstood. The standard is, in actuality, two separate standards. One regulates hazardous waste operations, §1910.120(a) through §1910.120(p). The other, §1910.120(q), is designed to control spills and releases of any hazardous substance **no matter where it occurs**. If either of those matters is a concern in your business, familiarize yourself with the applicable provisions of that standard.

The §1910.120 standard has 4 explanatory appendices (A through D) that can be very helpful to understanding its requirements. If you still have questions after reading them, contact your state consultative service.

The §1910.119 **process safety management** standard, only applies to processes that involve some 125 particular chemicals (listed by name and CAS number in Appendix A of §1910.119) when used in quantities exceeding specified thresholds listed in that Appendix, and processes that involve a flammable liquid or gas (as defined in §1910.1200(c)) on site in one location in a quantity of 10,000 pounds or more. Section 1910.119 does not apply to oil or gas well drilling or servicing operations, retail facilities, or normally unoccupied remote facilities.

7.   **Subpart I. Personal Protective Equipment**

SUBPART I - PERSONAL PROTECTIVE EQUIPMENT	
1910.132	General requirements
1910.133	Eye and face protection
1910.134	Respiratory protection
1910.135	Occupational head protection
1910.136	Occupational foot protection
1910.137	Electrical protective equipment
1910.138	Hand protection
1910.139	Sources of standards
1910.140	Standards organizations

The first three standards in Subpart I are among the most frequently cited by OSHA's inspectors. Consequently, if you have employees that work where personal protective equipment is (or may be) needed, you should make sure that the standards' requirements are observed.

OSHA has revised portions of the general industry safety standards addressing personal protective equipment (PPE). The revised standards include those containing general requirements for all PPE (1910.132) and standards that meet design, selection, and use requirements for specific types of PPE (eye, face, head, foot, and hand.)

New paragraphs (d), (e), and (f) (containing requirements covering equipment selection, defective and damaged equipment, and training, respectively) have been added to 1910.132. Also a new section (1910.138) has been added to this subpart to address hazards to the hands.

The new PPE standard has been interpreted to cover everything that could conceivably be regarded as "personal protective equipment" and every situation that could even remotely be deemed "hazardous."

§1910.133 covers eye and face protection and requires each affected employee must use appropriate eye or face protection when exposed to eye or face hazards from flying particles, molten metal, liquid chemicals, acids or caustics fluids, chemical gases or vapors, or potentially injurious light radiation.

- Detachable side protectors (e.g. clip-on or slide-on side shields) meeting the pertinent requirements of this section are acceptable.

- The Compliance Safety Health Officer (CSHO) shall assure that each employee who wears prescription lenses while engaged in operations that involve eye hazards is wearing eye protection that incorporates the prescription in its design, or protection that can be worn over the prescriptive lenses without disturbing the proper position of the prescriptive lenses.

- The CSHO shall assure that each affected employee using protective eyewear with filter lenses has eyewear with a shade number appropriate for the work being performed for protection from injurious light radiation.

§1910.135 covers Head Protection and requires each affected employee to wear protective helmets when working in areas where there is potential for injury to the head from falling objects. Also, protective helmets shall be designed to reduce electrical shock hazards when employees are working near exposed electrical conductors.

§1910.136 covers Foot Protection and requires each affected employee to wear protective footwear when working in areas where there is a danger of foot injuries due to falling and rolling objects, objects piercing the sole, and where an employee's feet are exposed to electrical hazards.

§1910.137 covers Electrical Protective Equipment and requires each affected employee to use electrical protective equipment such as insulating blankets, matting, covers, line hose, gloves, and sleeves made of rubber when exposed to electrical hazards.

§1910.138 covers Hand Protection and requires employees to use appropriate hand protection when employee's hands are exposed to hazards such as those from skin absorption of harmful substances; severe cuts or lacerations; severe abrasions; punctures; chemical burns; thermal burns; and harmful temperature extremes.

The personal protective standards 1910.132 through 1910.138 establish the employer's obligation to provide PPE to employees. The standard requires employers to provide and pay for PPE required by the company for the worker to do his/her job safely in compliance with OSHA standards.

If you have a respirator or gas mask on your premises -- even if no one ever uses it -- you must comply with §1910.134. It requires, among other things, that you establish and maintain a respiratory protective program. That means you must have your own written program. There are many additional details in §1910.134. They are summarized on the following chart.

MINIMAL ACCEPTABLE RESPIRATOR PROGRAM

Requirement	Standard
Written Operating Procedures	.134(b)(1), (e)(1), and (e)(3)
Proper Selection	.134(b)(2), (c), and (e)(2)
Training and Fitting	.134(b)(3), (e)(5), and (e)(5)(i-iii)
Cleaning and Disinfecting	.134(b)(5) and (f)(3)
Storage	.134(b)(6), and (f)(5)(i-iii)
Inspection and Maintenance	.134(b)(7), (e)(4), (f)(2)(i-iv), and (f)(4)
Work Area Surveillance	.134(b)(8) only
Inspection/Evaluation of Program	.134(b)(9) only
Medical Examinations	.134(b)(10) only
Approved Respirators	.134(b)(11) only

The other 3 Subpart I standards (§§1910.135, 1910.136 and 1910.137) simply reference the specifications to which hard hats, safety shoes and electrical personal protective equipment must conform. They don't impose any "use" or "provision" obligations on employers.

8.   Subpart J. General Environmental Controls

SUBPART J - GENERAL ENVIRONMENTAL CONTROLS	
1910.141	Sanitation
1910.142	Temporary labor camps
1910.144	Safety color code for marking physical hazards
1910.145	Accident prevention signs and tags
1910.146	Permit-required confined spaces
1910.147	Control of hazardous energy (lockout/tagout)
1910.148	Standards organizations
1910.149	Effective dates
1910.150	Sources of standards

This Subpart contains only 6 standards, one of which applies only to places that are used as Temporary Labor Camps, §1910.142.

§1910.141 imposes rather general requirements for housekeeping, as well as lavatory and toilet facilities at all permanent places of employment. You ought to familiarize yourself with that standard.

§1910.144 and §1910.145 cover the specifications for accident prevention signs and tags, and the **color codes** that are to be used on safety cans, emergency stop bars on machines, and physical hazards that can result in an employee tripping, falling, stumbling, striking against or being caught in-between.

§1910.146 went into effect on April 15, 1993. It applies to each place of employment where there is a "confined space." That term is defined as a space that:

(1)   Is large enough and so configured that an employee can bodily enter and perform assigned work; and

(2)   Has limited or restricted means for entry or exit (for example, tanks, vessels, silos, storage bins, hoppers, vaults, and pits are spaces that may have limited means of entry); and

(3)   Is not designed for continuous employee occupancy.

§1910.146(b). If you have any such space, you must study and observe §1910.146. It is quite detailed and it includes a number of paperwork requirements.

The lockout/tagout standard, §1910.147, is widespread in its application, detailed in its requirements, and frequently cited by OSHA inspectors. If you have machines in your business that are ever repaired, serviced or maintained, you must become familiar with §1910.147. Among other things, it requires that you adopt a written program setting forth the means and methods that are used in your business to observe the standard's lockout/tagout requirements, provide employees with particular kinds of training, and keep records of various activities that the standard requires.

9.   Subpart K. Medical and First Aid

SUBPART K - MEDICAL AND FIRST AID	
1910.151	Medical services and first aid
1910.153	Sources of standards

This Subpart is one of the briefest of all "general industry" subparts. It requires that you (1) have a person with first aid training on hand *unless* there is a hospital, clinic or infirmary "in near proximity," (2) ensure the "ready availability" of medical personnel, and (3) have "suitable facilities" for quick drenching or flushing of the eyes and body whenever your employees could be exposed to "injurious corrosive materials."

OSHA inspectors cite the eye-flushing standard, §1910.151(c), quite often -- and they frequently claim that the term "suitable facilities" means a particular kind of eyewash station. The case law, however, is otherwise. It holds that lavatories, sinks, drinking fountains and water hoses can be "suitable facilities" for the drenching or flushing of the eyes. If an employer has such a source of running water, OSHA must prove that it is **not** suitable in order to prove a §1910.151(c) violation. *Secretary of Labor v. Trinity Industries, Inc.*, 15 BNA OSHC 1985, 1987 (1992).

10. Subpart L. Fire Protection

SUBPART L - FIRE PROTECTION	
1910.155	Scope, application and definitions applicable to this subpart
1910.156	Fire brigades
PORTABLE FIRE SUPPRESSION EQUIPMENT	
1910.157	Portable fire extinguishers
1910.158	Standpipe and hose systems
FIXED FIRE SUPPRESSION EQUIPMENT	
1910.159	Automatic sprinkler systems
1910.160	Fixed extinguishing systems, general
1910.161	Fixed extinguishing systems, dry chemical
1910.162	Fixed extinguishing systems, gaseous agent
1910.163	Fixed extinguishing systems, water spray and foam
OTHER FIRE PROTECTION SYSTEMS	
1910.164	Fire detection systems
1910.165	Employee alarm systems

151

This Subpart contains requirements for fire brigades, and for all portable and fixed fire suppression equipment, fire detection systems, and fire and employee alarm systems that are installed in order to meet the OSHA fire protection requirements.

It does *not* require an employer to organize a fire brigade but, if he chooses to do so, it imposes numerous requirements that must be observed. They are explained rather clearly in Appendix A which follows §1910.165 in the Code of Federal Regulations.

If you have any of the other equipment and systems listed in the titles of the various standards included in Subpart L, you should acquaint yourself with the relevant standards. The easiest way to do that is to read Appendix A mentioned above. It contains a separate explanation of each standard. There are 4 other appendices that follow Appendix A, each of which could be quite helpful to understanding your Subpart L obligations.

Section 1910.157(g) requires that each employer must provide an educational program to familiarize employees with (1) the general principles of fire extinguisher use, and (2) the hazards involved with incipient-stage fire fighting, **if** the employer provides portable fire extinguishers for employee use. A program of that kind is not required where employees are **not** expected to fight fires. A chart that diagrams that requirement appears on the next page.

You should keep in mind that all OSHA fire protection requirements are not included in Subpart L. Fire Exits, fire prevention plans, emergency action plans, and Means of Egress are regulated by Subpart E. They were discussed earlier. Consequently, to make sure you are in compliance with all OSHA fire protection rules, you must familiarize yourself with both Subpart E and those provisions of Subpart L that are applicable to your business.

If you have portable fire extinguishers on the premises, the nature of your OSHA obligations depends upon whether or not your employees will use those extinguishers to fight fires. The following is a graphic depiction of that OSHA requirement.

FIGHT-NO FIGHT DECISION TREE FOR COMPLIANCE WITH SUBPART "L"

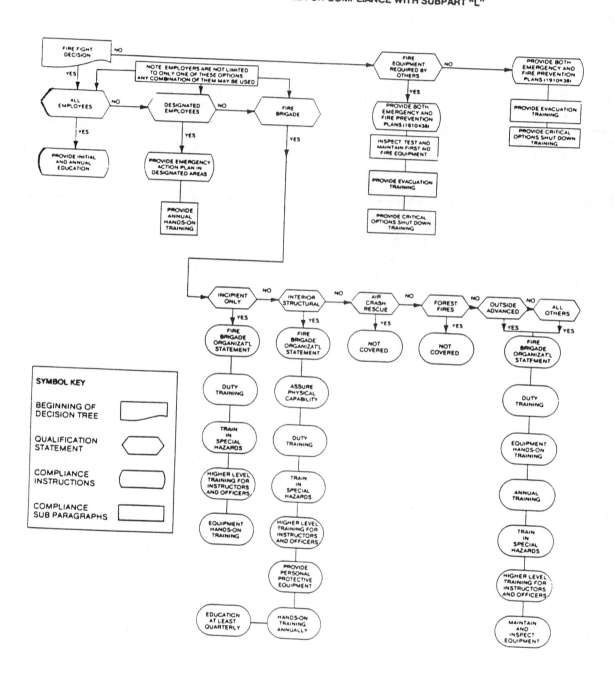

Of course, if you have any questions or need explanations, the experts at your state consultative service are there to help you. Check the listing for your state in part three of this book.

11. **Subpart M. Compressed Gas/Compressed Air Equipment**

SUBPART M - COMPRESSED GAS AND COMPRESSED AIR EQUIPMENT	
1910.169	Air receivers
1910.170	Sources of standards
1910.171	Standards organizations

This Subpart contains only one standard, §1910.169. It only applies to compressed air receivers, and other equipment that is used in providing and utilizing compressed air for performing operations such as cleaning, drilling, hoisting, and chipping. If you have or use equipment of that kind, you must aquaint yourself with §1910.169.

The standard does *not* cover the use of compressed air to convey materials, nor the problems created when employees work **in** compressed air (as in tunnels and caissons). Nor does it apply to compressed air machinery and equipment used on transportation vehicles such as steam railroad cars, electric railway cars, and automotive equipment.

Employers who have compressed air on the premises should also be familiar with §1910.242(b), discussed later under Subpart P. It prohibits use of compressed air for any cleaning purpose *unless* reduced to 30 p.s.i.

12. Subpart N. Materials Handling and Storage

SUBPART N - MATERIALS HANDLING AND STORAGE	
1910.176	Handling materials - general
1910.177	Servicing multi-piece and single piece rim wheels
1910.178	Powered industrial trucks
1910.179	Overhead and gantry cranes
1910.180	Crawler locomotive and truck cranes
1910.181	Derricks
1910.182	Effective dates
1910.183	Helicopters
1910.184	Slings
1910.189	Sources of standards
1910.190	Standards organizations

This Subpart is very limited in its application. It covers only the moving and storage of materials, the servicing of rim wheels used on large vehicles (trucks, tractors, buses, etc.), and the use of the particular items of equipment that are listed in the title of each standard.

If any of that applies in your business, familiarize yourself with the general requirements set forth in §1910.176 **and** the particular requirements for those items of equipment mentioned above that you use in your business operations.

§1910.177 requires training for all tire servicing employees that work on trucks, tractors, trailers, buses and off-road vehicles. It requires the utilization of industry-accepted procedures that minimize the potential for employee injury, the use of proper equipment such as clip-on chucks, restraining devices or barriers to retain the wheel components in the event of an incident during the inflation of tires, and the use of compatible components. The standard does *not* apply to the servicing of rim wheels used on automobiles, or on pickup trucks and vans that utilize automobile or truck tires designated "LT."

Section 1910.178 contains safety requirements for fire protection, design, maintenance, and use of fork lift trucks, tractors, platform lift trucks, motorized hand trucks, and other specialized industrial trucks that are powered by electric motors or internal combustion engines. It does *not* apply to compressed air or nonflammable compressed gas-operated industrial trucks, nor to farm vehicles, nor to vehicles intended primarily for earth moving or over-the-road hauling.

Section 1910.179 applies to overhead and gantry cranes, including semi-gantry, cantilever gantry, wall cranes, storage bridge cranes, and others having the same fundamental characteristics.

Section 1910.180 applies to crawler cranes, locomotive cranes, wheel-mounted cranes of both truck and self-propelled wheel type, and any variations thereof which retain the same fundamental characteristics. This standard only covers cranes of the above types, which are basically powered by internal combustion engines or electric motors and which utilize drums and ropes. Cranes designed for railway and automobile wreck clearances are excepted. The requirements of this standard are applicable only to machines when used as lifting cranes.

Section 1910.181 (Derricks) applies to guy, stiffleg, basket, breast, gin pole, Chicago boom, and A-frame derricks of the stationary type, capable of handling loads at variable reaches and powered by hoists through systems of rope reeving, used to perform lifting hook work, single or multiple line bucket work, grab, grapple, and magnet work.

The term "derrick" is defined as an apparatus consisting of a mast or equivalent member held at the head by guys or braces, with or without a boom, for use with a hoisting mechanism and operating ropes. Derricks may be permanently installed for temporary use as in construction work. This standard also applies to any modification of the types of derrick mentioned above which retain their fundamental features, except for floating derricks.

§1910.183 applies only to helicopter crane operations.

§1910.184 applies where slings are used in conjunction with other material handling equipment for the movement of material by hoisting. The types of slings covered are those made from alloy steel chain, wire rope, metal mesh, natural or synthetic fiber rope (conventional three strand construction), and synthetic web (nylon, polyester, and polypropylene).

13. Subpart O. Machinery and Machine Guarding

SUBPART O - MACHINERY AND MACHINE GUARDING	
1910.211	Definitions
1910.212	General requirements for all machines
1910.213	Woodworking machinery requirements
1910.214	Cooperage machinery
1910.215	Abrasive wheel machinery
1910.216	Mills and calendars in the rubber and plastics industries
1910.217	Mechanical power presses
1910.218	Forging machines
1910.219	Mechanical power-transmission apparatus
1910.220	Effective dates
1910.221	Sources of standards
1910.222	Standards organizations

Subpart O contains specific requirements for the 7 particular kinds of machinery and apparatus listed above in the titles for §1910.213 through §1910.219. If your business uses those things, familiarize yourself with the relevant standard or standards.

Three of the specifically-applicable Subpart O standards (§1910.215, §1910.216 and §1910.219) need a little more explanation.

Section 1910.215 only covers *abrasive* wheel machinery. It does not cover wire wheels, buffing wheels or the like. An abrasive wheel is made up of individual particles that are bonded together to form a wheel. The hazard sought to be controlled by the standard is that, if not properly mounted and used, the wheel can literally explode. Sections of the wheel may fly out at high speeds and can strike the operator causing death or serious injury. If abrasive wheels are used in hand or portable power tools, the Subpart P standards (discussed later) should be consulted.

Section 1910.216 regulates mills and calendars only when used in the rubber and plastics industries. The term "mill" is defined in §1910.211(c)(3).

Section 1910.219 contains detailed requirements for the safeguarding of the components of the mechanical power transmission system which transmit energy from the prime mover (power source) to the part of the machine performing the work. Those components include flywheels, pulleys, belts, connecting rods, shafting, couplings, cams, spindles, chains, cranks, and gears. The primary thrust of §1910.219 is to ensure that employees cannot be injured from being caught by rotating members, in-running nip points, sprockets or pulleys, and the like.

Subpart O also contains general guarding requirements for all machines. Those requirements are contained in §1910.212. That is not only one of the most cited OSHA standard but it is worded so broadly that it can have many different meanings. OSHA interprets it to mean that: 1.) Any machine part, function, or process which may cause injury must be safeguarded, and 2.) Where the operation of a machine or accidental contact with it can injure the operator or others in the vicinity, the hazard must be either controlled or eliminated. The courts, however, have consistently held that §1910.212 does not apply unless there exists an actual hazard to an employee - not a conceivable one.

To prove a §1910.212(a)(1) violation, OSHA must do more than show that it **may** be physically possible for an employee to come into contact with the unguarded machinery in questions. OSHA must establish that employees are exposed to hazard as a result of the manner in which the machine functions and the way it is operated. *Secretary of Labor v. Jefferson Smurfit Co.*, 15 BNA OSHC 1419, 1421 (1991).

If any standard included in Subpart O applies to your operations, you must familiarize yourself with it. You should also read the definition of terms that are included in §1910.211.

Keep in mind that the maintenance servicing and repair of machinery is regulated by §1910.147, a standard (frequently called the "lockout/tagout" standard) that is part of Subpart J (discussed earlier).

If you still have doubts or questions, contact your state consultative service. See Part Three of this book.

14.   Subpart P. Hand Tools/Portable Power Tools/Compressed Air

SUBPART P - HAND AND PORTABLE POWERED TOOLS AND OTHER HAND-HELD EQUIPMENT	
1910.241	Definitions
1910.242	Hand and portable powered tools and equipment, general
1910.243	Guarding of portable powered tools
1910.244	Other portable tools and equipment
1910.245	Effective dates
1910.246	Sources of standards
1910.247	Standards organizations

If your business uses compressed air or portable tools -- whether powered or non-powered (i.e., hand tools) -- you should peruse the Subpart P standards. They cover less than 9 pages in the Code of Federal Regulations.

The title used in Subpart P is somewhat misleading. It covers much more than its title suggests. For example, the title contains no reference to the most-cited OSHA standard in this subpart, §1910.242(b). That standard requires that:

"Compressed air shall not be used for cleaning purposes except where reduced to less than 30 p.s.i. and then only with effective chip guarding and personal protective equipment."

Although Subpart P contains some rather general requirements for all hand tools and power tools and equipment, it also includes specific provisions for many others, including the following:

- Power Lawnmowers
- Drills
- Saws
- Sanders
- Grinders
- Shears
- Planers
- Jacks
- Trimmers
- Abrasive Wheels
- Grinding Wheels
- Explosive Actuated Fastening Tools
- Abrasive Blast Cleaning Nozzles

If you use any of those in your business, consult the relevant portions of Subpart P, particularly §1910.243 and §1910.244.

15.   Subpart Q. Welding, Cutting and Brazing

SUBPART Q - WELDING, CUTTING AND BRAZING	
1910.251	Definitions
1910.252	General requirements
1910.253	Oxygen-fuel gas welding and cutting
1910.254	Arc welding and cutting
1910.255	Resistance welding
1910.256	Sources of standards
1910.257	Standards organizations

Subpart Q regulates electric and gas welding, cutting and brazing. Employers who have those kinds of operations in their business should become familiar with the OSHA standards included in Subpart Q and, if compressed gases are used in those operations, Subpart H as well.

Many welding and cutting operations require the use of compressed gases. When compressed gases are consumed in the welding process, such as oxygen-fuel gas welding, the requirements for their handling, storage, and use are contained in Subpart Q. However, general requirements for the handling, storage, and use of compressed gases are contained in Subpart H-Hazardous Materials, §§1910.101 through 1910.105.

Certain welding and cutting operations require the use of compressed gases other than those consumed in the welding process. For example, gas metal arc welding utilizes compressed gases for shielding. Handling, storage, and use of compressed gases in situations such as those requires compliance with the requirements contained in Subpart H. See the discussion of Subpart H earlier in this section of the book.

The "general requirements" for welding, cutting and brazing that are covered in §1910.252 actually contain detailed and rather specific rules for **Fire prevention** and protection during those operations, §1910.252(a), **Protection of personnel**, §1910.252(b), **Health protection and ventilation** (particularly when substances such as fluorine compounds, lead, beryllium, cadmium, cleaning compounds and stainless steels are involved), §1910.252(c), and **Industrial applications**, §1910.252(d).

The remaining 3 standards in Subpart Q are restricted to the particular operations listed below:

Oxygen-fuel gas welding and cutting. §1910.253.

Arc welding and cutting. §1910.254.

Resistance welding. §1910.255.

Employers with operations of the kind discussed or mentioned above must familiarize themselves with the relevant standards.

16.   **Subpart R. Special Industries.**

SUBPART R - SPECIAL INDUSTRIES	
1910.261	Pulp, paper, and paperboard mills
1910.262	Textiles
1910.263	Bakery equipment
1910.264	Laundry machinery and operations
1910.265	Sawmills
1910.266	Pulpwood logging
1910.268	Telecommunications
1910.269	Electric power generation, transmission, and distribution
1910.272	Grain handling facilities
1910.274	Sources of standards

The standards contained in Subpart R are different from those in the other Part 1910 subparts because the application of each of them is limited to **one particular industry**. If you are engaged in one of the nine industries listed in the Subpart R standards, then you must familiarize yourself with the relevant standard and observe it. If you are not engaged in one of those industries, then you can ignore the Subpart R standards.

Employers engaged in the 9 Subpart R industries should be cautioned that the particular standard that applies to their industry does not include all their OSHA obligations. Those industries also must observe all applicable OSHA standards **but** a specifically applicable Subpart R standard may pre-empt other OSHA standards.

There is an OSHA interpretative rule that explains the foregoing, §1910.5. Among other things, that rule states that: "If a particular standard is specifically applicable to a condition, practice, means, method, operation, or process, it shall prevail over any different general standard which might otherwise be applicable to the same condition, practice, means, method, operation, or process." For example, if you run a bakery which has machines to handle manually fed dough, there exists a specific machine guarding standard, §1910.263(h). That preempts OSHA's general machine guarding standard, §1910.212.

The rule then goes on to state: "On the other hand, any standard shall apply according to its terms to any employment and place of employment in any industry, even though particular standards are also prescribed for the industry, as in Subpart R, to the extent that none of such particular standards applies. To illustrate, the general standard regarding noise exposure in §1910.95 applies to employments and places of employment in pulp, paper, and paperboard mills covered by §1910.261."

17.   **Subpart S. Electrical**

SUBPART S - ELECTRICAL	
GENERAL	
1910.301	Introduction
DESIGN SAFETY STANDARDS FOR ELECTRICAL SYSTEMS	
1910.302	Electric utilization systems
1910.303	General requirements
1910.304	Wiring design and protection
1910.305	Wiring methods, components, and equipment for general use
1910.306	Specific purpose equipment and installations
1910.307	Hazardous (classified) locations
1910.308	Special systems
SAFETY-RELATED WORK PRACTICES	
1910.331	Scope
1910.332	Training
1910.333	Selection and use of work practices
1910.334	Use of equipment
1910.335	Safeguards for personnel protection
DEFINITIONS	
1910.399	Definitions applicable to this subpart

The Subpart S standards were extracted from the National Electrical Code (NEC), a privately-written code of conduct in general use for a hundred years or so. OSHA has taken the position that employers can consult the NEC for specific information on how the required performance can be obtained.

The Subpart S standards are grouped into two categories, as follows:

1.   Design of Electrical Systems: §1910.302 to §1910.308

2.   Safety-Related Work Practices: §1910.331 to §1910.335

Included in category 1 are the regulations for all electrical equipment and installations that are used to provide electrical power and light for places, buildings, structures and other premises where employees work, including yards, industrial substations, conductors that connect the installations to a supply of electricity, other outside conductors on the premises, carnivals, parking lots, mobile homes and recreational vehicles. That does not include automobiles, ships, railway rolling stock, aircraft, underground mines, equipment used to operate railways, telecommunication and electrical utilities.

Included in category 2 (above) are the regulations covering the practices and procedures that must be observed in order to protect employees who are working on or near exposed energized and deenergized parts of electric equipment. They include a requirement that each employer with employees who do that kind of work must have a written program that describes the means and methods to be used in observing the standard's requirements. §1910.333(b)(2)(i). Those means and methods are set forth rather explicitly in §1910.333(b). Indeed, OSHA has taken the position that the standard's written program requirement can be satisfied if the employer has on hand a copy of §1910.333(b).

Employee training requirements are also included in Subpart S. All employees who face a risk of electric shock, burns or other related injuries that are not reduced to a safe level by the *installation* safety requirements of Subpart S, must be trained in the safety-related *work practices* required by §1910.331 through §1910.335.

In addition to being trained in, and familiar with, safety related work practices, some employees must also be trained in the inherent hazards of electricity, such as high voltages, electric current, arcing, grounding, and lack of guarding. Any electrically related safety practices not specifically addressed by Sections 1910.331 through 1910.335, but necessary for safety in specific workplace conditions -- must be included in the training.

Employers who desire additional explanation of the Subpart S installation requirements might want to obtain a series of 16 fact sheets that OSHA has developed for that purpose. They include helpful checklists, "do's and don't's" and graphic representations. All 16 come in a package of 58 typewritten pages (reprinted January 1992) entitled: "OSHA Electrical Hazard Fact Sheet" and include references to the corresponding section of the National Electrical Code. A list of their contents follows:

18.   **Subpart T. Commercial Diving Operations**

SUBPART T - COMMERCIAL DIVING OPERATIONS	
GENERAL	
1910.401	Scope and application
1910.402	Definitions
PERSONNEL REQUIREMENTS	
1910.410	Qualifications of dive team
GENERAL OPERATIONS PROCEDURES	
1910.420	Safe practices manual
1910.421	Pre-dive procedures
1910.422	Procedures during dive
1910.423	Post-dive procedures
SPECIFIC OPERATIONS PROCEDURES	
1910.424	SCUBA diving
1910.425	Surface-supplied air diving
1910.426	Mixed-gas diving
1910.427	Liveboating
EQUIPMENT PROCEDURES AND REQUIREMENTS	
1910.430	Equipment
RECORDKEEPING	
1910.440	Recordkeeping
1910.441	Effective date

As its title makes clear, Subpart T applies only to commercial diving operations. Employers engaged in those operations must read and observe the standards included in Subpart T. All other employers can ignore it.

Although it is included in the Part 1910 "General Industry" standards, Subpart T applies to diving and related support operations conducted in connection with all types of work and employments, including general industry, construction, ship repairing, shipbuilding, shipbreaking and longshoring. However, it does not apply to "scientific diving" (see Appendix B of the standard for an explanation) or to any diving operation:

(i) Performed solely for instructional purposes, using open-circuit, compressed-air SCUBA and conducted within the no-decompression limits;

(ii) Performed solely for search, rescue, or related public safety purposes by or under the control of a governmental agency; or

(iii) Governed by 45 CFR Part 47 (Protection of Human Subjects, U.S. Department of Health and Human Services) or equivalent rules or regulations established by another federal agency, which regulate research, development, or related purposes involving human subjects.

19.   **Subpart Z. Toxic and Hazardous Substances**

SUBPART Z - TOXIC AND HAZARDOUS SUBSTANCES	
1910.1000	Air contaminants
1910.1001	Asbestos
1910.1002	Coal tar pitch volatiles; interpretation of term
1910.1003	4-Nitrobiphenyl
1910.1004	Alpha-Naphthylamine
1910.1004	Methyl chloromethyl ether
1910.1004	3,3' - Dichlorobenzidine (and its salts)
1910.1004	Bis-Chloromethyl ether
1910.1004	Beta-Naphthylamine
1910.1004	Benzidine
1910.1004	4-Aminodiphenyl
1910.1004	Ethyleneimine
1910.1004	Beta-Propiolactone
1910.1004	2-Acetylaminofluorene
1910.1004	4-Dimethylaminoazobenzene
1910.1004	N-Nitrosodimethylamine
1910.1017	Vinyl chloride
1910.1018	Inorganic arsenic
1910.1025	Lead

Continued on next page

Subpart Z (Continued)

1910.1027	Cadmium
1910.1028	Benzene
1910.1029	Coke oven emissions
1910.1030	Bloodborne Pathogens
1910.1043	Cotton dust
1910.1044	1,2-dibromo-3-chloropropane
1910.1045	Acrylonitrile
1910.1047	Ethylene oxide
1910.1048	Formaldehyde
1910.1050	Methylenedianiline
1910.1096	Ionizing radiation
1910.1200	Hazard communication
1910.1201	Retention of DOT markings, placards, and labels
1910.1450	Occupational exposure to hazardous chemicals in laboratories

The Subpart Z standards set airborne permissible exposure limits (PELs) for over 450 listed substances, impose a number of specific requirements to be implemented in workplaces where those substances are present, and require employers, chemical manufacturers and importers, to provide information upon **all** hazardous chemicals through a variety of specified methods.

A PEL is the maximum amount of a contaminant in the air to which workers may be exposed over a given period of time.

§1910.1000 lists 428 substances in three different tables (Z-1, Z-2 and Z-3) that literally cover the alphabet from Acetaldehyde to Zirconium and sets a PEL for each of them. Employers are required to keep those substances within the PEL if it is feasible to do so or, if not, require exposed employees to use personal protective equipment that keeps their exposure within the listed PEL.

§1910.1001 through §1910.1050 are limited to the single substance listed in each standard's title. Those standards impose additional requirements that are not included in §1910.1000. For example, adoption of a written compliance program for each workplace where the substance is present, periodic monitoring of the workplace air to measure contaminant levels, employee information and training, medical monitoring of each employee exposed to the substance, transfer or removal of employees under certain circumstances, and various other requirements.

If any of your employees could be exposed to any of the substances covered by the standards in §1910.1001 through §1910.1050, you must familiarize yourself with the relevant standard and observe its requirements.

With your copy of the Part 1910 standards in hand, you should also read the list of substances that are included in Table Z-1, Z-2 and Z-3 of §1910.1000, determine whether any of them are present in your workplace and, if so, whether the airborne level of each such substance is within the specified PEL. If the listed PEL is *not* exceeded, you have no need to be concerned with §1910.1000. If it *is* exceeded, you must familiarize yourself with that standard and take the measures specified in §1910.1000.

Section 1910.1450 is limited in its application to laboratories where hazardous chemicals are handled or used on a "laboratory scale". Read the definition of that term in §1910.1450(b) to determine whether or not you are within the standard's scope. If so, read and observe the entire standard.

Every employer must pay strict attention to §1910.1200, the OSHA Hazard Communication standard. It requires that: **First**, chemical manufacturers and importers must assess the hazards of all chemicals that they produce or import and furnish detailed information to their customers upon those determined to be hazardous, and **second**, all employers must provide that information to their employees by means of a written hazard communication program, labels on containers, material safety data sheets, employee training and access to written records and documents.

The term "hazardous chemical" is defined very broadly (for example, table salt is included) so virtually every employer is required to observe §1910.1200. It is also the most cited OSHA standard.

There are numerous vendors and private consultants who can provide you with help in meeting your hazard communication obligations. OSHA and each state OSHA office has explanatory materials available for your use and the free consultative service that exists in each state stands ready to help you. Check the nearest office in the state-by-state listings in part three of this book, then give them a call.

§1910.1201 Retention of Department of Transportation (DOT), markings, placards, and labels. Each employer who receives a package of hazardous material which is required to be marked, labeled or placed in compliance with DOT's Hazardous materials regulations (49 CFR Parts 171-180) must keep those markings, labels, and placards on the package until the package is sufficiently cleaned of residue to remove any potential hazards.

IV.   OSHA MARITIME STANDARDS

1.   Overview

Unlike the Part 1910 "general industry" standards that are grouped within a single "part" of 29 CFR, the maritime standards have been placed in 4 separate parts -- even though there are fewer of them. The 4 parts, however, regulate separate and distinct employments, as listed below:

> PART 1915. Shipyard Employment[1]
> PART 1917. Marine Terminals
> PART 1918. Longshoring
> PART 1919. Gear Certification

Each of the four have a number of Subparts. They are discussed individually below.

None of the OSHA maritime standards apply to matters within the regulatory authority of the U.S. Coast Guard. In situations where an OSHA standard and a Coast Guard regulation overlap, the OSHA standard becomes unenforceable because of the existence of section 4(b)(1) of the OSH Act (Nothing in this Act shall apply to working conditions of employees with respect to which other Federal agencies exercise statutory authority to prescribe or enforce standards or regulations affecting occupational safety or health). 29 U.S.C. §653(b)(1).

---

[1]For its first 11 years of existence, the Part 1915 standards were separated into three groups, one each for ship repairing, shipbuilding and shipbreaking operations. Those three were consolidated into one (entitled "Shipyard Employment") in 1982. See 47 Federal Register 16984, April 20, 1982.

The Part 1915 Shipyard standards apply to all ship repairing, shipbuilding and shipbreaking operations and related employments.

The Part 1917 Marine Terminal standards apply at all wharves, bulkheads, quays, piers, docks and other berthing locations and adjacent storage or contiguous areas and structures associated with the primary movements of cargo or materials from vessel to shore or shore to vessel including structures which are devoted to receiving, handling, holding, consolidation and loading or delivery of waterborne shipments and passengers, as well as areas devoted to the maintenance of the terminal or equipment.

The marine terminal standards do *not* apply, however, to production or manufacturing areas having their own docking facilities and located at a marine terminal, to storage facilities directly associated with those production or manufacturing areas, to facilities used solely for the bulk storage, handling and transfer of flammable, non-flammable and combustible liquids and gases, to facilities subject to the regulations of the Office of Pipeline Safety Regulation of the U.S. Department of Transportation, or to fully automated bulk coal handling facilities contiguous to electrical power generating plants.

The Part 1918 Longshoring standards apply to the loading, unloading, moving, or handling of cargo, ship's stores, gear, etc., into, in, on, or out of any vessel on the navigable waters of the United States.

The Part 1919 Gear Certification standards are somewhat different. They set forth the OSHA procedures and standards that govern the accreditation of persons for the purpose of (1) Certificating vessels' cargo gear and shore-based material handling devices, and (2) The manner in which such certification shall be performed.

2.   Part 1915. Shipyard Standards

Part 1915 is divided into 12 Subparts each of which is designated by a capital letter and a title. They are as follows:

- Subpart A -   General Provisions
- Subpart B -   Explosive and Other Dangerous Atmospheres
- Subpart C -   Surface Preparation and Preservation
- Subpart D -   Welding, Cutting and Heating
- Subpart E -   Scaffolds, Ladders and Other Working Surfaces
- Subpart F -   General Working Conditions
- Subpart G -   Gear and Equipment for Rigging and Materials Handling
- Subpart H -   Tools and Related Equipment
- Subpart I -   Personal Protective Equipment
- Subpart J -   Ship's Machinery and Piping Systems
- Subpart K -   Portable, Unfired Pressure Vessels, Drums and Containers, Other Than Ship's Equipment
- Subpart L -   Electrical Machinery
- Subpart Z -   Toxic and Hazardous Substances

Each of those Subparts are further broken down into sections. Each will be listed individually by Subpart, section and title:

A. Subpart A. General Provisions

SUBPART A - GENERAL PROVISIONS	
1915.1	Purpose and authority
1915.2	Scope and application
1915.3	Responsibility
1915.4	Definitions
1915.5	Reference specifications, standards and codes
1915.6	Commercial diving operations
1915.7	Competent person

Subpart A defines some of the terms that are used in other subparts of the Part 1915 standards.

Section 1915.6 simply requires employers engaged in commercial diving operations to observe the standards (discussed earlier) that are included in Subpart T of Part 1910.

Section 1915.7 requires employers to designate certain employees as "competent persons" (and to list them on OSHA Form 73) in order to fulfill the requirements (contained in other Part 1915 subparts) to limit certain jobs, duties and responsibilities to "competent persons." That requirement does *not* apply, however, if those duties are always carried out by an NFPA Certified Marine Chemist.

That standard also establishes criteria to guide employers in making the required designation, §1915.7(b), and requires that written records must be made on OSHA Form 74 for each test and inspection that is required to be performed by any standard included in Subparts B, C, D and H of Part 1915, except the §1915.35(b)(8) and §1915.36(a)(5) standards in Subpart C.

B.   Subpart B.   Explosives/Dangerous Atmospheres

SUBPART B - EXPLOSIVE AND OTHER DANGEROUS ATMOSPHERES	
1915.11	Scope and application of subpart
1915.12	Precautions before entering
1915.13	Cleaning and other cold work
1915.14	Certification before hot work is begun
1915.15	Maintaining gas-free conditions
1915.16	Warning signs

The Subpart B standards are limited to work performed in cargo (or other) spaces that contain -- or having last contained -- combustible or flammable liquids or gases in bulk, as well as spaces that are immediately adjacent to them, on tank vessels, pipe lines, dry cargo, miscellaneous and passenger vessels. Each of the requirements specified in those standards must be observed whenever employees are assigned (or permitted) to do work in such places.

C.   Subpart C. Surface Preparation and Preservation

SUBPART C - SURFACE PREPARATION AND PRESERVATION	
1915.31	Scope and application of subpart
1915.32	Toxic cleaning solvents
1915.33	Chemical paint and preservative removers
1915.34	Mechanical paint removers
1915.35	Painting
1915.36	Flammable liquids

The Subpart C standards prescribe the precautions and employee protection requirements that must be observed when employees do the kind of jobs listed in the standards' titles, or work with the identified substances and materials.

D.   Subpart D. Welding, Cutting and Heating

WELDING, CUTTING AND HEATING	
1915.51	Ventilation and protection in welding, cutting and heating
1915.52	Fire prevention
1915.53	Welding, cutting and heating in way of preservative coatings
1915.54	Welding, cutting and heating of hollow metal containers and structures not covered by §1915.12
1915.55	Gas welding and cutting
1915.56	Arc welding and cutting
1915.57	Uses of fissionable material in ship repairing and shipbuilding

The Subpart D standards prescribe the precautions and employee protection requirements that must be observed when employees do welding, cutting or heating work, or engage in activities involving the use of and exposure to any source of ionizing radiation or radioactive materials.

E.   Subpart E. Scaffolds/Ladders/Working Surfaces

SUBPART E - SCAFFOLDS, LADDERS AND OTHER WORKING SURFACES	
1915.71	Scaffolds or staging
1915.72	Ladders
1915.73	Guarding of deck openings and edges
1915.74	Access to vessels
1915.75	Access to and guarding of dry docks and marine railways
1915.76	Access to cargo spaces and confined spaces
1915.77	Working surfaces

The Subpart E standards establish the requirements for use of ladders, scaffolds and similar apparatus; require the use of a gangway, ramp or permanent stairway when accessing vessels, dry docks and marine railways; provide for guardrails or the equivalent when employees work around openings or unguarded edges of decks, platforms and similar places; require the elimination of tripping hazards; and mandate use of personal

The **specifications** for certain ladders, scaffolds and related devices are set forth in Table E which is part of 29 C.F.R. §1915.118 (Subpart G).

F.   Subpart F. General Working Conditions

SUBPART F - GENERAL WORKING CONDITIONS	
1915.91	Housekeeping
1915.92	Illumination
1915.93	Utilities
1915.94	Work in confined or isolated spaces
1915.95	Ship repairing and shipbuilding work on or in the vicinity of radar and radio
1915.96	Work in or on lifeboats
1915.97	Health and sanitation
1915.98	First aid
1915.100	Retention of DOT markings, placards, and labels

The Subpart F standards provide for the maintenance of first aid kits, stokes basket stretchers and rules for the general working conditions that must be observed when employees work in or on lifeboats; in confined (or isolated) spaces; and on masts, king posts or other loft areas. It further requires clear walkways and working surfaces, adequate illumination of work stations, the keeping of flammable substances (and rags contaminated with oil, paint, etc) in fire resistant covered containers when not in use, and sets forth the rules that must be followed when connecting and using temporary lights, off-vessel steam supply, hoses and electric power, and for guarding infrared electrical heat lamps.

G.   Subpart G. Gear/Rigging & Materials Handling Equipment

SUBPART G - GEAR AND EQUIPMENT FOR RIGGING AND MATERIALS HANDLING	
1915.111	Inspection
1915.112	Ropes, chains and slings
1915.113	Shackles and hooks
1915.114	Chain falls and pull-lifts
1915.115	Hoisting and hauling equipment
1915.116	Use of gear
1915.117	Qualifications of operators
1915.118	Tables

Subpart G applies to the maintenance and use of the articles of equipment and apparatus listed in each standard's title and to the qualifications of the persons who are permitted to operate cranes, winches and all other power-operated hoisting apparatus. The **specifications** for the ropes, slings, chains, clips, shackles, horses, ladders and scaffolds, as well as the filter lenses for protection against radiant energy, appear in the tables that follow §1915.118.

H.   Subpart H. Tools and Related Equipment

SUBPART H - TOOLS AND EQUIPMENT	
1915.131	General precautions
1915.132	Portable electric tools
1915.133	Hand tools
1915.134	Abrasive wheels
1915.135	Power actuated fastening tools
1915.136	Internal combustion engines other than ship's equipment

Subpart H imposes requirements for the items of equipment identified in the title to each of its standards as well as precautions for using them. It also provides for guarding and securing of moving parts, as well as the labelling and inspection of compressed air hoses and lines.

I.   Subpart I. Personal Protective Equipment

SUBPART I - PERSONAL PROTECTIVE EQUIPMENT	
1915.151	Scope, application, and definitions
1915.152	General requirements
1915.153	Eye and face protection
1915.154	Respiratory protection
1915.155	Head protection
1915.156	Foot protection
1915.157	Hand and body protection
1915.158	Lifesaving equipment
1915.159	Personal fall arrest systems
1915.160	Positioning device systems

Subpart I sets forth the conditions that require employee use of the various types of personal protective equipment identified in the standards' titles (including safety belts, lifelines and life rings), the specifications for such equipment, and for its inspection, maintenance, storage and use. It requires the use of respirators for employee protection when working in confined spaces and places where there are air contaminants.

J.   Subpart J. Ship's Machinery and Piping Systems

SUBPART J - SHIP'S MACHINERY AND PIPING SYSTEMS	
1915.161	Scope and application of subpart
1915.162	Ship's boilers
1915.163	Ship's piping systems
1915.164	Ship's propulsion machinery
1915.165	Ship's deck machinery

Subpart J imposes requirements that must be observed when employees work on or in the listed machinery, systems and apparatus identified in the standard's titles.

K.   Subpart K. Portable Air Receivers/Unfired Pressure Vessels

SUBPART K - PORTABLE, UNFIRED PRESSURE VESSELS, DRUMS AND CONTAINERS, OTHER THAN SHIP'S EQUIPMENT	
1915.171	Scope and application of subpart
1915.172	Portable air receivers and other unfired pressure vessels
1915.173	Drums and containers

Subpart K regulates the inspection, use, storage, and specifications for, the identified equipment and apparatus.

L.   Subpart L.   Electrical Machinery

SUBPART L - ELECTRICAL MACHINERY	
1915.181	Electrical circuits and distribution boards

Subpart L contains only one standard. It requires the deenergizing of electrical circuits whenever electrical work is done and imposes various requirements that must be observed when working on electrical circuits, energized parts, and energized boards.

M.   Subpart Z.   Toxic and Hazardous Substances

Subpart Z is **identical** to Subpart Z of Part 1910 of the General Industry Standards discussed earlier. The only difference is the **numbers** used to identify the standards. The first four digits of each is "1915" instead of "1910". For example, §1910.1000 of the General Industry standards is designated as §1915.1000 in the Shipyard Standards. Shipyard employers should read and observe the discussion of the Subpart Z, Part 1910, standards earlier in this book.

Subpart Z also includes §1915.1120, the standard providing for Access to Employee Exposure and Medical Records. That standard is identical to the "General Industry" standard, §1910.20, which was discussed earlier in this book under "OSHA Recordkeeping and Reporting Regulations."

3.   Part 1917. Marine Terminals

Part 1917 applies to Marine Terminals as described above at page. It is divided into the following 7 subparts:

- Subpart A - Scope and Definitions

- Subpart B - Marine Terminal Operations

- Subpart C - Cargo Handling Gear and Equipment

- Subpart D - Specialized Terminals

- Subpart E - Personal Protection

- Subpart F - Terminal Facilities

- Subpart G - Related Terminal Operations and Equipment

Subpart A defines a number of the terms used in Part 1917 and relates its scope and applicability (see above for that).

The remaining subparts (B through G) provide the requirements that must be observed for the matters identified in the title for each standard in that subpart set forth below:

SUBPART B - MARINE TERMINAL OPERATIONS

1917.11	Housekeeping
1917.12	Slippery conditions
1917.13	Slinging
1917.14	Stacking of cargo and pallets
1917.15	Coopering
1917.16	Line handling
1917.17	Railroad facilities
1917.18	Log handling
1917.19	Movement of barges and rail cars
1917.20	Interference with communications
1917.21	Open fires
1917.22	Hazardous cargo (see §1917.2(p))
1917.23	Hazardous atmospheres and substances (see §1917.2(p))
1917.24	Carbon monoxide
1917.25	Fumigants, pesticides, insecticides and hazardous preservatives (see §1917.2(p))
1917.26	First aid and lifesaving facilities
1917.27	Personnel
1917.28	Hazard communication[2]

---

[2]The hazard communication standard is identical to the "General Industry" standard. That standard was discussed earlier in this book.

SUBPART C - CARGO HANDLING GEAR AND EQUIPMENT	
1917.41	House falls
1917.42	Miscellaneous auxiliary gear
1917.43	Powered industrial trucks
1917.44	General rules applicable to vehicles
1917.45	Cranes and derricks (see also §1917.51)
1917.46	Crane load and limit devices
1917.47	Winches
1917.48	Conveyors
1917.49	Spouts, chutes, hoppers, bins, and associated equipment
1917.50	Certification of marine terminal material handling devices
1917.51	Hand tools

SUBPART D - SPECIALIZED TERMINALS	
1917.70	General
1917.71	Terminals handling intermodal container or roll-on roll-off operations
1917.73	Terminal facilities handling menhaden and similar species of fish

SUBPART E - PERSONAL PROTECTION	
1917.91	Eye protection
1917.92	Respiratory protection
1917.93	Head protection
1917.94	Foot protection
1917.95	Other protective measures

SUBPART F - TERMINAL FACILITIES

1917.111	Maintenance and load limits
1917.112	Guarding of edges
1917.113	Clearance heights
1917.114	Cargo doors
1917.115	Platforms and skids
1917.116	Elevators and escalators
1917.117	Manlifts
1917.118	Fixed ladders
1917.119	Portable ladders
1917.120	Fixed stairways
1917.121	Spiral stairways
1917.122	Employee exits
1917.123	Illumination
1917.124	Passage between levels and across openings
1917.125	Guarding temporary hazards
1917.126	River banks
1917.127	Sanitation
1917.128	Signs and marking

SUBPART G - RELATED TERMINAL OPERATIONS AND EQUIPMENT	
1917.151	Machine guarding
1917.152	Welding, cutting and heating (hot work)
1917.153	Spray painting
1917.154	Compressed air
1917.155	Air receivers
1917.156	Fuel handling and storage
1917.157	Battery charging and changing
1917.158	Prohibited operations

As a general rule, OSHA's Part 1910 "General Industry" standards do **not** apply at marine terminals. However, §1917.1(a)(2) identifies the following Part 1910 standards that marine terminal employers must observe:

1. *Safety requirements for scaffolding*, Subpart D, §1910.28.
2. *Abrasive blasting.* Subpart G, §1910.94(a).
3. *Noise.* Subpart G, §1910.95.
4. *Respiratory protection.* Subpart I, §1910.134.
5. *Servicing multi-piece and single piece rim wheels.* Subpart N, §1910.177.
6. *Grain handling facilities.* Subpart R, §1910.272.
7. *Electrical.* Subpart S.
8. *Toxic and hazardous substances.* Subpart Z applies where specifically referenced in Part 1917, except that the requirements of Subpart Z do **not** apply when a substance or cargo is contained within a sealed, intact means of packaging or containment complying with Department of Transportation or International Maritime Organization requirements.
9. *Commercial Diving Operations.* Subpart T.
10. *Access to employee exposure and medical records.* Subpart C, §1910.20.

4.   Part 1918. Longshoring

Part 1918 applies to Longshoring. It consists of 10 subparts as listed below:

- Subpart A - General Provisions
- Subpart B - Gangways and Gear Certification
- Subpart C - Means of Access
- Subpart D - Working Surfaces
- Subpart E - Opening and Closing Hatches
- Subpart F - Ship's Cargo Handling Gear
- Subpart G - Cargo Handling Gear and Equipment Other Than Ship's Gear
- Subpart H - Handling Cargo
- Subpart I - General Working Conditions
- Subpart J - Personal Protective Equipment

Subpart A defines a number of terms used in Part 1918 and states that the 1968

ANSI (American National Standards Institute) standard for eye and face protection

(Z87.1 - 1968) and the 1969 ANSI head protection standard (Z89.1 - 1969) are

incorporated by reference. Longshoring employers must therefore observe those

standards. For the most part, however, those two standards include requirements similar

to those in the Part 1910 "General Industry" standards (§1910.133 and §1910.135) that

were described earlier in the Subpart I discussion of this book. Section 1918.6(b) states

that copies of the standards may be obtained from any local or regional OSHA office (see

the list in part three of this book).

The remaining subparts of Part 1918 set forth the requirements that must be

observed for the matters identified in the title for each standard included in that subpart.

They are listed below:

SUBPART B - GANGWAYS AND GEAR CERTIFICATION	
1918.11	Gangways
1918.12	Gear certification
1918.13	Certification of shore-based material handling devices
1918.14	Container cranes
1918.15	Effective date of §§1918.13 and 1918.14.

SUBPART C - MEANS OF ACCESS	
1918.21	Gangways and other means of access
1918.22	Jacob's ladders
1918.23	Access of barges and river towboats
1918.24	Bridge plates and ramps
1918.25	Ladders

SUBPART D - WORKING SURFACES	
1918.31	Hatch coverings
1918.32	Stowed cargo and temporary landing platforms
1918.33	Deck loads
1918.34	Skeleton decks
1918.35	Open hatches
1918.36	Weather deck rails
1918.37	Barges
1918.38	Freshly oiled decks

SUBPART E - OPENING AND CLOSING HATCHES	
1918.41	Coaming clearances
1918.42	Beam and pontoon bridles
1918.43	Handling beams and covers

SUBPART F - SHIP'S CARGO HANDLING GEAR	
1918.51	General requirements
1918.52	Specific requirements
1918.53	Cargo winches
1918.54	Rigging gear
1918.55	Cranes

SUBPART G - CARGO HANDLING GEAR	
1918.61	General
1918.62	Fiber rope and fiber rope slings
1918.63	Wire rope and wire rope slings
1918.64	Chains and chain slings
1918.65	Shackles
1918.66	Hooks other than hand hooks
1918.67	Pallets
1918.68	Chutes, gravity conveyors and rollers
1918.69	Powered conveyors

SUBPART G - CARGO HANDLING GEAR	
1918.73	Mechanically-powered vehicles used aboard vessels
1918.74	Cranes and derricks other than vessel's gear
1918.75	Notifying ship's officers before using certain equipment
1918.76	Grounding

SUBPART H - HANDLING CARGO	
1918.81	Slinging
1918.82	Building drafts
1918.83	Stowed cargo, tiering and breaking down
1918.84	Bulling cargo
1918.85	Containerized cargo
1918.86	Hazardous cargo

SUBPART I - GENERAL WORKING CONDITIONS	
1918.90	Hazard communication[3]
1918.91	Housekeeping
1918.92	Illumination
1918.93	Ventilation and atmospheric conditions
1918.94	Sanitation and drinking water
1918.95	Longshoring operations in the vicinity of repair and maintenance work
1918.96	First aid and life saving equipment
1918.97	Qualification of machinery operators
1918.98	Grain fitting
1918.99	Commercial diving operations
1918.100	Retention of DOT markings, placards, and labels

SUBPART J - PERSONAL PROTECTIVE EQUIPMENT	
1918.101	Eye protection
1918.102	Respiratory protection
1918.103	Protective clothing
1918.104	Foot protection
1918.105	Head protection
1918.106	Protection against drowning

[3]The hazard communication standard is identical to the "General Industry" standard. It was discussed earlier in this book.

5.   Part 1919. Gear Certification

Part 1919 "Gear Certification" standards cover the OSHA procedures and standards governing the accreditation of persons who are authorized to certificate vessels' cargo gear and shore-based material handling devices, as well as the manner in which certification shall be performed.

A vessel's cargo gear documentation consists of a booklet called a "Register" which has four or five parts (the fifth is optional). Entries are made in the register and cover annual and quadrennial surveys of the assembled gear and its components. The type of entries depend upon how the vessel is equipped. Additional entries may be made with respect to certain items or equipment, such as validation of performance of a repair or replacement of some component, and in many cases a supporting certificate may be required.

The Register also contains instructions outlining the applicable requirements and a notation as to who may conduct the surveys. Appropriate unit and loose gear test certificates containing the applicable test results must be kept with the Register. The documentation used by all Maritime nations is essentially the same. Foreign documentation with some exceptions is printed in English, or English and the appropriate foreign language.

A cargo gear Register with supporting documents is not issued when shore-based material-handling devices are certificated. However, documents attesting to the certification performed by accredited persons are required by the OSHA standards in 29 CFR §§1915.115(a)(1), 1918.13, 1918.14 and 1926.605(a). Those documents, must be on the original government form as follows:

(1)   OSHA-71 Form (LBS-00S-MAR-14), "Certificate of Unit Test and/or Examination of Crane, Derrick or other Material-Handling Devices."

(2)   OSHA-72 Form (LBS-00S-MAR-15), "Notice to Owners of Deficiencies Found on Certification Survey."

The Part 1919 standards should be read in conjunction with Longshoring standards §§1918.12(c), 1918.12(d)(3) and 1918.13 because they implement those standards. Part 1919 has 8 subparts, as follows:

- Subpart A - General Provisions
- Subpart B - Procedure Governing Accreditation
- Subpart C - Duties of Persons Accredited to Certificate Vessels' Cargo Gear
- Subpart D - Certification of Vessels Cargo Gear
- Subpart E - Certification of Vessels; Tests and Proof Loads; Heat Treatment;   Competent Persons
- Subpart F - Accreditation to Certificate Shore-Based Equipment
- Subpart G - Duties of Persons Accredited to Certificate Shore-Based Material Handling Devices
- Subpart H - Certification of Shore-Based Material Handling Devices

Subpart A defines some of the terms used in Part 1919 and sets forth some of the circumstances under which accreditation is **not** required. The remaining subparts state the requirements to be observed for the matters identified in the title for each standard

included in that subject. They are listed below:

SUBPART B - PROCEDURE GOVERNING ACCREDITATION	
1919.3	Application for accreditation
1919.4	Action upon application
1919.5	Duration and renewal of accreditation
1919.6	Criteria governing accreditation to certificate vessels' cargo gear
1919.7	Voluntary amendment or termination of accreditation
1919.8	Suspension or revocation of accreditation
1919.9	Reconsideration and review

SUBPART C - DUTIES OF PERSONS ACCREDITED TO CERTIFICATE VESSELS' CARGO GEAR	
1919.10	General duties; exemptions
1919.11	Recordkeeping and related procedures concerning records in custody of accredited persons
1919.12	Recordkeeping and related procedures concerning records in custody of the vessel

SUBPART D - CERTIFICATION OF VESSELS' CARGO GEAR	
1919.13	General
1919.14	Initial tests of cargo gear and tests after alterations, renewals or repairs
1919.15	Periodic tests, examinations and inspections
1919.16	Heat treatment
1919.17	Exemptions from heat treatment
1919.18	Grace periods
1919.19	Gear requiring welding
1919.20	Damaged components
1919.21	Marking and posting of safe working loads
1919.22	Requirements governing braking devices and power sources
1919.23	Means of derrick attachment
1919.24	Limitations on use of wire rope
1919.25	Limitations on use of chains

SUBPART E - CERTIFICATION OF VESSELS; TESTS AND PROOF LOADS; HEAT TREATMENT; COMPETENT PERSONS

1919.26	Visual inspection before tests
1919.27	Unit proof tests - winches, derricks and gear accessory thereto
1919.28	Unit proof tests - cranes and gear accessory thereto
1919.29	Limitations on safe working loads and proof loads
1919.30	Examinations subsequent to unit tests
1919.31	Proof tests - loose gear
1919.32	Specially designed blocks and components
1919.33	Proof tests - wire rope
1919.34	Proof tests after repairs or alternations
1919.35	Order of tests
1919.36	Heat treatment
1919.37	Competent persons

SUBPART F - ACCREDITATION TO CERTIFICATE SHORE-BASED EQUIPMENT

1919.50	Eligibility for accreditation to certificate shore-based material handling devices covered by §1918.13 of the safety and health regulations for long-shoring
1919.51	Provisions respecting application for accreditation, action upon the application, and related matters

SUBPART G - DUTIES OF PERSONS	
1919.60	General duties, exemptions

SUBPART H - CERTIFICATION OF SHORE-BASED MATERIAL HANDLING DEVICES	
1919.70	General provisions
1919.71	Unit proof test and examination of cranes
1919.72	Annual examination of cranes
1919.73	Unit proof test and examination of derricks
1919.74	Annual examination of derricks
1919.75	Determination of crane or derrick safe working loads and limitations in absence of manufacturer's data
1919.76	Safe working load reduction
1919.77	Safe working load increase
1919.78	Nondestructive examinations
1919.79	Wire rope
1919.80	Heat treatment
1919.81	Examination of bulk cargo loading or discharging spouts or suckers
1919.90	Documentation

V.   OSHA CONSTRUCTION INDUSTRY STANDARDS

1.   Overview

Construction safety standards were initially adopted by the U.S. Department of Labor in April 1971 to implement the Contract Work Hours and Safety Standards Act, 40 U.S.C. §333, also known as the Construction Safety Act. Shortly thereafter, they were converted into OSHA standards under the authority that Congress delegated to the Secretary of Labor in section 6(a) of the OSH Act.

An employer must comply with the safety and health regulations in Part 1926 if its employees are "engaged in construction work". See: 29 C.F.R. §1910.12(b). In order to determine if such work, defined as "work for construction, alteration, and/or repair, including painting and decorating," is being performed, one must consult 29 C.F.R. §1926.13 for a discussion of those terms.

Section 1926.13 states:

The terms "construction," "alteration," and "repair" used in section 107 of the [Construction Safety] Act are also used in section 1 of the Davis-Bacon Act (40 U.S.C. 276a), providing minimum wage protection on Federal construction contracts, and section 1 of the Miller Act (40 U.S.C. 270a), providing performance and payment bond protection on Federal construction contracts....*The use of the same or identical terms in these statutes which apply concurrently with section*

107 of the [Construction Safety] Act have considerable precedential value in
ascertaining the coverage of section 107.

29 C.F.R. §1926.13(a) (emphasis added).

Thus, §1926.13(a) requires that the terms "construction, alteration, and repair" in
the Construction Safety Act be interpreted consistently with those same terms as used in
the Davis-Bacon Act and the Miller Act. Under Labor Department regulations
implementing the Davis-Bacon Act:

The terms "construction"... or "repair" mean all types of work done on a particular
building or work *at the site* thereof... all work done in the construction or
development of the project, including without limitation, altering, remodeling,
installation (where appropriate) *on the site of the work* of items fabricated off-site,
painting and decorating, the transporting of materials and supplies to or from the
building or work... and the manufacturing or furnishing of materials, articles,
supplies or equipment *on the site of the building* or work...

29 C.F.R. §5.2(j) (emphasis added). "The 'site of the work' is limited to the physical
place or places where the construction called for in the contract will remain when work
on it has been completed..." 29 C.F.R. §5.2(l)(1).

There has been some litigation in the past over the meaning of the term "engaged
in construction work." One of the more extensive discussions of that question is

contained in a 1987 U.S. Court of Appeals decision: *Brock v. Cardinal Industries, Inc.,* 828 F.2d 373 (6th Cir. 1987).

The Part 1926 standards include 26 subparts (subpart A through subpart Z) but subparts A and B apply only to determining the scope of section 107 of the Construction Safety Act, 40 U.S.C. §333. That Act applies only to employers who are engaged in construction under contract with the U.S. government. OSHA does not base citations upon either Subpart A or B. Consequently, no further consideration will be given to them in this book. The remaining 24 subparts are listed below:

- Subpart C - General Safety and Health Provisions
- Subpart D - Occupational Health and Environmental Controls
- Subpart E - Personal Protective and Life Saving Equipment
- Subpart F - Fire Protection and Prevention
- Subpart G - Signs, Signals, and Barricades
- Subpart H - Materials Handling, Storage, Use, and Disposal
- Subpart I - Tools - Hand and Power
- Subpart J - Welding and Cutting
- Subpart K - Electrical
- Subpart L - Scaffolding
- Subpart M - Fall protection
- Subpart N - Cranes, Derricks, Hoists, Elevators, and Conveyors
- Subpart O - Motor Vehicles, Mechanized Equipment, and Marine Operations
- Subpart P - Excavations
- Subpart Q - Concrete and Masonry Construction
- Subpart R - Steel Erection
- Subpart S - Underground Construction, Caissons, Cofferdams, and Compressed Air
- Subpart T - Demolition
- Subpart U - Blasting and the Use of Explosives
- Subpart V - Power Transmission and Distribution
- Subpart W- Rollover Protective Structures; Overhead Protection

- Subpart X - Stairways and Ladders
- Subpart Y - Commercial Diving Operations
- Subpart Z - Toxic and Hazardous Substances

Employers engaged in construction should read over each of those subparts and identify those that could be applicable to their own operations, then read the discussion of those subparts that follows.

OSHA has also identified a number of Part 1910 "General Industry" standards that apply to construction work. In June 1993, each of them was assigned a "Part 1926" designation and re-issued as part of the construction standards. Thus, they will be included in the discussion that follows.

2. Subpart C. General Safety and Health Provisions

SUBPART C - GENERAL SAFETY AND HEALTH PROVISIONS	
1926.20	General safety and health provisions
1926.21	Safety training and education
1926.23	First aid and medical attention
1926.24	Fire protection and prevention
1926.25	Housekeeping
1926.26	Illumination
1926.27	Sanitation
1926.28	Personal protective equipment
1926.29	Acceptable certifications
1926.30	Shipbuilding and ship repairing
1926.31	Incorporation by reference
1926.32	Definitions
1926.33	Acess to employee exposure and medical records
1926.34	Means of egress
1926.35	Employee emergency action plans

The "general" provisions included in §1926.20 require construction employers to:

(1) Initiate and maintain such **programs** as may be necessary to comply with the Part 1926 standards, and to include as a requirement of those programs a provision requiring that frequent and regular inspections be made by "competent persons" of the job sites, materials and equipment,

(2) Identify (as unsafe) and prohibit the use of: machinery, tools, materials and equipment that is not in compliance with any Part 1926 standard, and

(3) Limit the operation of equipment and machinery to only those employees who are qualified by training or experience to operate it.

Some OSHA inspectors insist that §1926.20(b) requires that each construction employer must adopt a written accident prevention program for each job site. The standard, however, does not include the word "written". Employers should pay close attention to §1926.20(b). It is one of the most cited construction safety standards.

Section 1926.21 requires that employees be given instruction (not training) in the recognition and avoidance of unsafe conditions and the regulations applicable (to his work) regarding hazard control, as well as the necessary precautions and the necessary protective equipment for those employees who are required to enter "confined or enclosed spaces." Although §1926.21 is somewhat ambiguous, it is a favorite of OSHA inspectors. Innumerable citations have been issued under that standard.

The standards in §§1926.23 through 1926.28 apply to most construction jobs and are stated succinctly and briefly. Each employer should familiarize himself with them.

The "acceptable certifications" standard in §1926.29 applies only to boilers, pressure vessels and similar equipment.

The provisions of §1926.30 and §1926.31 do not impose any substantive requirements on employers.

Section 1926.32 contain definitions for 18 terms that are used in Part 1926, including "competent person", "employee", "employer", "qualified", "safety factor", "shall", "should", and "suitable".

The three remaining standards are "General Industry" standards that OSHA has incorporated into the construction standards. §1926.33 is the same as §1910.20 (discussed earlier in this book under "OSHA Recordkeeping and Reporting Regulations). §1926.34 requires that there shall be free egress from every building and structure. It is essentially the same as Part 1910, Subpart E, Means of Egress. See the discussion of that topic earlier in this book.

§1926.35 is the same as §1910.38, a standard that is also part of Subpart E, Part 1910, Means of Egress.

3. Subpart D. Occupational Health and Environmental Controls

SUBPART D - OCCUPATIONAL HEALTH & ENVIRONMENT CONTROLS	
1926.51	Sanitation
1926.52	Occupational noise exposure
1926.53	Ionizing radiation
1926.54	Nonionizing radiation
1926.55	Gases, vapors, fumes, dusts, and mists
1926.56	Illumination
1926.57	Ventilation
1926.59	Hazard communication[4]
1926.60	Methylenedianiline (MDA)
1926.61	Retention of DOT markings, placards, and labels
1926.62	Lead
1926.64	Process safety management of highly hazardous chemicals.
1926.65	Hazardous waste operations and emergency response
1926.66	Criteria for design and construction for spray booths

---

[4] The hazard communication standard is identical to the "General Industry" standard.

There are 7 Part 1910 "general industry" standards that apply to Subpart D. They are:

§1910.141(a)(1). Scope of general industry sanitation standards
§1910.141(a)(2)(v). "Potable water" defined
§1910.141(a)(5). Vermin control
§1910.141(g)(2). Eating and drinking areas
§1910.141(h). Food handling
§1910.151(c). Medical services and first aid - quick drenching facilities
§1910.161(a)(2). Carbon dioxide extinguishing systems, safety requirements

The medical services, sanitation and illumination requirements in §§1926.50, 1926.51 and 1926.56 apply at all construction sites.

§1926.52 is the noise standard for the construction industry. It limits employee noise exposure to 90 dBA averaged over an 8-hour day but, unlike the general industry noise standard, it does *not* require the same kind of hearing conservation program (monitoring, audiometric testing, etc.). If you have employees exposed to high noise levels, you should also observe §1926.102 (part of Subpart E) which requires ear protective devices. For all intents and purposes, compliance with §1926.102 will satisfy the construction industry hearing conservation requirement.

The §1926.53 standard applies to activities that involve ionizing radiation, radio-active materials or x-rays. The §1926.54 standard applies when laser equipment (nonionizing radiation) is used. The use of laser safety goggles is required by a subpart E standard, §1926.102(b)(2).

§1926.55 is the construction equivalent of the Part 1910, Subpart "Z", general industry standards. As originally issued, it did not list the 400-odd regulated substances by name or the employee exposure limitations for those substances. It simply stated that

ACGIH (American Conference of Governmental Industrial Hygienists) "Threshold Limit

Values of Airborne Contaminants for 1970"[5] shall be avoided. If not, controls must first

be implemented whenever feasible. When not feasible, personal protective equipment

must be used. Those ACGIH limits are now listed in Appendix A of §1926.55.

Construction employers should read the list of substances that appear in the standard's

Appendix A. If the limits listed there are exceeded, the §1926.55 requirements must be

observed.

§1926.56 sets forth specific illumination intensities for construction areas, ramps,

runways, corridors, offices, shops and storage areas.

If ventilation is used as an engineering control to limit exposure to airborne

contaminants, the §1926.57 "ventilation" standard applies.

If your work involves exposure to MPD, you must be familiar with §1926.60

and carefully observe its requirements. It includes innumerable details. This is also true

for employers whose work involves exposure to lead. They must observe the numerous

requirements included in this standard. §1926.62 (lead)

Retention of Department of Transportation (DOT) markings, placards, and labels.

§1926.61 Each employer who receives a package of hazardous material which is required

to be marked, labeled or placarded in compliance with DOT's Hazardous Materials

Regulations (49 CFR Parts 171 through 180) must keep those markings, labels and

---

[5]For the most part, the particular substances regulated by the 1970 ACGIH threshold limit values and the airborne exposure limit for each is the same as the PELs listed in Tables Z-1, Z-2 and Z-3 of the §1910.1000 general industry standard.

placards on the package until the packaging is sufficiently cleaned of residue to remove any potential hazards.

The markings, placards, and labels must be maintained in a manner that ensures that they are readily visible.

Process Safety Management of highly hazardous chemicals §1926.64 is the same as the part 1910 "General Industry" standard. See the discussion earlier in this book under the heading "Subpart H. Hazardous Materials."

Hazardous waste operators and Emergency Response §1926.65 is the same as the part 1910 "General Industry" standard. See the discussion earlier in this book under the heading "Subpart H. Hazardous Materials."

Criteria for design and construction for spray booths §1922.66 incorporates into the construction standards the General Industry's requirements §1910.107 discussed earlier in this book under the heading "Subpart H. Hazardous Materials."

The Hazard Communication standard is the #1 most cited OSHA standard. Every employer must pay strict attention to §1926.59. It requires that first: chemical manufacturers and importers must assess the hazards of all chemicals that they produce or import and furnish detailed material safety data sheets (MSD's) to their customers upon those determined to be hazardous. Second, all employers must provide that information to their employees by means of a **written** Hazard Communication program, labels on containers, MSDS sheets, employee training, and employees must have access to written records of all of this.

The term "Hazardous Chemical" is defined very broadly (for example, table salt is included). So virtually every employee is required to observe §1926.59.

OSHA and each state OSHA office has explanatory materials available for your use and the free consultative service that exists in each state is always ready to help you. Check the nearest office in the state-by-state listings in part three of this book, then give them a call.

4.   Subpart E. Personal Protective and Life Saving Equipment

SUBPART E - PERSONAL PROTECTIVE AND LIFE SAVING EQUIPMENT*	
1926.56	Criteria for personal protective equipment
1926.96	Occupational foot protection
1926.100	Head protection
1926.101	Hearing protection
1926.102	Eye and face protection
1926.103	Respiratory protection
1926.104	Safety belts, lifelines, and lanyards
1926.105	Safety nets
1926.106	Working over or near water
1926.107	Definitions applicable to this subpart

The Subpart E standards state in very general terms, the conditions under which the protective equipment identified in the standards' titles must be used. §1926.95 is the same as §1910.132 discussed earlier in this book under the heading: "Subpart I. Personal Protective Equipment." §1926.96 simply sets forth the specifications for safety-toe footwear. §1926.105 provides for safety nets for elevated workplaces 25 feet or more above the adjoining surface it is impractical to use ladders, scaffolds, temporary floors, or safety lines or belts as protection against falls.

§1926.107 defines 6 of the terms used in Subpart E:

       (a). Contaminant

       (b). Lanyard

       (c). Lifeline

       (d). O.D. (optical density)

       (e). Radiant Energy

       (f). Safety Belt

5.   Subpart F. Fire Protection and Prevention

SUBPART F - FIRE PROTECTION AND PREVENTION[6]	
1926.150	**Fire protection**
1926.151	**Fire prevention**
1926.152	**Flammable and combustible liquids**
1926.153	**Liquefied petroleum gas (LP-Gas)**
1926.154	**Temporary heating devices**
1926.155	**Definitions applicable to this subpart**

The fire protection and fire prevention requirements in §§1926.150 and 1926.151 are rather detailed and should be read and observed by all construction employers. Among other things, those standards require the development of a fire protection program, §1926.150(a), a trained fire brigade "as warranted by the project", §1926.150(a)(5), and periodic inspection/maintenance of all portable fire extinguishers, §1926.150(c)(1)(viii).

If flammable or combustible liquids (or LP Gas) are present at a construction site, the detailed requirements of §1926.152 (or §1926.153) must be observed.

§1926.154 sets forth the rules for the use of temporary heating devices.

---

[6] OSHA has identified parts of 9 general industry standards that are applicable to Subpart F. They are:

Service station class I liquid storage tanks inventory records, §1910.106(g)(1)(i)(g),
Marine service station dispensing, §1910.106(g)(4)(i) through §1910.106(g)(4)(iii),
Scope of general industry flammable liquids standards, §1910.106(j)(1) and §1910.106(j)(2),
Definition of "Marine service station", §1910.106(a)(22),
Liquefied petroleum gas containers defined, §1910.110(a)(1) through §1910.110(a)(4),
LP-gas container labeling, §1910.110(b)(5)(iii),
Systems using containers other than DOT containers, §1910.110(d)(1) and §1910(d)(2),
Installation of storage containers, §1910.110(d)(7)(vii) and §1910.110(d)(7)(viii), and
Damage from vehicles, §1910.110(d)(10).

SUBPART G - SIGNS, SIGNALS, AND BARRICADES	
1926.200	Accident prevention signs and tags
1926.201	Signaling
1926.202	Barricades
1926.203	Definitions applicable to this subpart

Subpart G is very brief. It requires the posting of construction areas with legible traffic signs at points of hazard, §1926.200(g), and the use of flagmen or other appropriate traffic controls when the necessary protection cannot be provided by signs, signals and barricades. §1926.201. The other Subpart G "standards" simply provide the specifications and color coding for signs and tags if they are used or required by some other OSHA standard.

7.   **Subpart H. Materials, Handling, Storage, Use, and Disposal**

SUBPART H - MATERIALS HANDLING, STORAGE, USE, AND DISPOSAL[7]	
1926.250	General requirements for storage
1926.251	Rigging equipment for material handling
1926.252	Disposal of waste materials

Section 1926.250 provides detailed requirements for the movement, stacking and storage of various building materials.

§1926.251 sets forth the requirements to be observed whenever rigging equipment is used in material handling.

The disposal of ordinary waste materials (*not* hazardous waste) at a construction site is regulated by §1926.252.

---

[7]OSHA has identified parts of 8 general industry standards as applicable to Subpart H. They are:

Dockboards, §1910.30(a)(1) through §1910.30(a)(5),
Housekeeping, Storage Areas, §1910.176(c),
Alloy steel chain slings, §1910.184(e)(3)(i) and §1910.184(e)(3)(ii),
Sling safe operating procedures, §1910.184(c)(2) through §1910.184(c)(12),
Wire rope slings, §1910.184(f)(2) through §1910.184(f)(4),
Natural and synthetic fiber rope slings, §1910.184(i)(2) through §1910.184(h)(5),
Synthetic web slings, §1910.184(i)(2) through §1910.184(i)(9), and
Scope of sling standards, §1910.184(a).

8.   Subpart I. Tools -- Hand and Power

SUBPART I - TOOLS - HAND AND POWER	
1926.300	General requirements
1926.301	Hand tools
1926.302	Power operated hand tools
1926.303	Abrasive wheels and tools
1926.304	Woodworking tools
1926.305	Jacks - lever and ratchet, screw and hydraulic[8]
1926.306	Air Receivers
1926.307	Mechanical power-transmission apparatus

The Subpart I standards set forth the requirements to be observed whenever the identified kinds of tools and equipment are used in construction. The "general requirements" standard, §1926.300, requires maintenance "in a safe condition" of all hand tools, power tools and similar equipment, the guarding of moving parts of equipment, and imposes requirements for use of personal protective equipment (when using hand and power tools) and "on-off" controls on power tools.

§1926.306 is the same as §1910.169. See the discussion earlier in this book under: "Subpart M. Compressed Gas and Compressed Air Equipment."

§1926.307 is the same as §1910.219. See the discussion earlier in this book under: "Subpart O. Machinery and Machine Guarding."

---

[8]Whenever hydraulic jacks are used in lift-slab operations, the requirements of §1926.705 (Subpart Q) must also be observed.

OSHA has also identified parts of 4 "general industry" standards that apply to Subpart I:

> Machine point of operations guarding, §1910.212(a)(3) and §1910.212(a)(5),
> Anchoring fixed machinery, §1910.212(b),
> Abrasive blast cleaning nozzles, §1910.244(b), and
> Jacks, operation and maintenance, §1910.244(a)(2)(iii) through
> §1910.244(a)(2)(viii).

9.   Subpart J. Welding and Cutting

SUBPART J - WELDING AND CUTTING	
1926.350	Gas welding and cutting
1926.351	Arc welding and cutting
1926.352	Fire prevention
1926.353	Ventilation and protection in welding, cutting, and heating
1926.354	Welding, cutting and heating in way of preservative coatings

Subpart J regulates all aspects of welding, cutting and heating when those operations are performed on a construction project. Employers should pay careful attention to those five Subpart J standards whenever welding or cutting is done.

10. Subpart K. Electrical

SUBPART K - ELECTRICAL	
GENERAL	
1926.400	Introduction
INSTALLATION SAFETY REQUIREMENTS	
1926.402	Applicability
1926.403	General requirements
1926.404	Wiring design and protection
1926.405	Wiring methods, components, and equipment for general use
1926.406	Specific purpose equipment and installations
1926.407	Hazardous (classified) locations
1926.408	Special systems
SAFETY-RELATED WORK PRACTICES	
1926.416	General requirements
1926.417	Lockout and tagging of circuits
SAFETY-RELATED MAINTENANCE AND ENVIRONMENTAL CONSIDERATIONS	
1926.431	Maintenance of equipment
1926.432	Environmental deterioration of equipment
SAFETY REQUIREMENTS FOR SPECIAL	
1926.441	Battery locations and battery charging
DEFINITIONS	
1926.449	Definitions applicable to this subpart

Subpart K covers the electrical safety requirements for construction jobsites. They are separated into four major divisions as indicated above (electrical equipment and installations, work practices, equipment maintenance/deterioration, and the "special" rules for batteries and battery charging). The electrical standards are quite detailed, so the definitions in §1926.449

should be consulted when reading them.

Section 1926.400 simply states how the electrical standards are arranged. Those

responsible for construction jobsites should familiarize themselves with the remaining provisions

of Subpart K. If there are doubts or questions, the consultative service for your state is there to

provide you with explanations. See Part Three of this book.

11.   Subpart L. Scaffolding

SUBPART L - SCAFFOLDING	
1926.450	Scope, application and definitions applicable to this subpart
1926.451	General requirements
1926.452	Requirements for specific types of scaffolds
1926.453	Aerial lifts
1926.454	Training requirements

OSHA has revised the scaffolding standard. It will allow employers greater flexibility in

the use of fall protection systems to protect employees working on scaffolds.

Employers who use scaffolds in construction work should be familiar with all the

definitions included in §1926.450. §1926.451 contains all the general requirements for all

scaffolds. It also sets the minimum strength criteria for all scaffold components and connections.

It requires that each scaffold component be capable of supporting without failure, its own weight

and at least 4 times the maximum intended load applied to it. §1926.452 has specific

requirements for 25 particular types of scaffolds, as follows:

Pole	§1926.452 (a)
Tube and Coupler	§1926.452 (b)
Fabricated Frame	§1926.452 (c)
Plasterers, decorators, and large area scaffolds	§1926.452 (d)
Bricklayers square	§1926.452 (e)
Horse	§1926.452 (f)
Form and carpenters	§1926.452 (g)
Roof bracket	§1926.452 (h)
Outrigger	§1926.452 (i)
Pump jack	§1926.452 (j)
Ladder jack	§1926.452 (k)
Window jack Chairs	§1926.452 (l)
Crawling boards	§1926.452 (m)
Step platform and trestle ladder	§1926.452 (n)
Single-point adjustable	§1926.452 (o)
Two-point adjustable suspension (Swing stages)	§1926.452 (p)
Multi-point suspension	§1926.452 (q)
Catenary	§1926.452 (r)
Float or ship	§1926.452 (s)
Interior hung	§1926.452 (t)
Needle beam	§1926.452 (u)
Multi-level	§1926.452 (v)
Mobile	§1926.452 (w)
Repair bracket	§1926.452 (x)
Stilts	§1926.452 (y)

§1926.453 is a new addition to the scaffold standard. It sets some specific requirements for lift operations. This will apply to equipment for vehicle mounted elevating and rotating work platforms.

§1926.454 adds a new section on training requirements. The standard distinguishes between the training needed by employees to erect and **to dismantle** scaffolds. The standard applies to all construction work. It requires employers to instruct each employee in the recognition and avoidance of unsafe conditions. It also sets certain criteria allowing employers to tailor training to fit the particular circumstances of each employer's workplace.

12. Subpart M. Fall Protection

SUBPART M - FALL PROTECTION	
1926.500	Scope, application, and definitions applicable to this subpart
1926.501	Duty to have fall protection
1926.502	Fall protection systems
1926.503	Training requirements

The Fall Protection standard contains OSHA's new revised regulations in construction workplaces to prevent employees from falling off, onto, or through levels, and to protect employees from being struck by falling objects.

Under the new standard, employers will be able to select fall protection measures compatible with the type of work being performed. Fall protection generally can be provided through the use of guardrail systems, safety net systems, personal fall arrest systems, positioning device systems, and warning line systems, among others.

§1926.501 requires employers to assess the workplace to determine if the walking/working surfaces on which employees are to work have the strength and structural integrity to safely support workers. Employees are not permitted to work on those surfaces until it has been determined that the surfaces have the requisite strength and structural integrity to support the workers. Once employers have determined that the surface is safe for employees to work on, the employer must select one of the options listed for the work operation if a fall hazard is present.

For example, if an employee is exposed to falling 6 feet or more from an unprotected side or edge, the employer must select either a guardrail system, safety net system, or personal fall arrest system to protect the worker.

§1926.502 Fall protection systems and criteria can be provided through the use of any of the following: guardrail systems, safety net systems or personal fall arrest system. However, under certain conditions, employers that cannot use conventional fall protection equipment must develop and implement a written fall protection plan specifying alternative fall protection measures.

If the employer chooses to use guardrail systems to protect workers from falls, the system must meet the following criteria: guardrails require a top rail around 42 inches, a midrail at around 21 inches and toe boards when necessary to prevent objects from falling over the edge.

If an employer chooses to use safety net systems, the system must be installed as close as practicable under the walking/working surface on which employees are working and never more than 30 feet (9.1 meters) below such levels. Defective nets shall not be used. Safety nets shall be inspected at least once a week for wear, damage, and other deterioration. The maximum size of each safety net mesh opening shall nto exceed 36 square inches (230 square centimeters) nor be longer than 6 inches (15 centimeters) on any side, and the openings, measured center-to-center, of mesh ropes or webbing, shall not exceed 6 inches (15 centimeters). All mesh crossings shall be secured to prevent enlargement of the mesh opening.

Safety nets shall be capable of absorbing an impact force of a drop test consisting of a 400-pound (180 kilogram) bag of sand 30 inches (76 centimeters) in diameter dropped from the highest walking/working surface at which workers are exposed, but not from less than 42 inches (1.1 meters) above that level.

If an employer chooses to use a personal fall arrest system, it will consist of an anchorage, connectors, and a body belt or body harness and may include a deceleration device, lifeline, or suitable combinations. If a personal fall arrest system is used for fall protection, it must do the following:

- Limit maximum arresting force on an employee to 900 pounds (4 kilonewtons) when used with a body belt;
- Limit maximum arresting force on an employee to 1,800 pounds (8 kilonewtons) when used with a body harness;
- Be rigged so that an employee can neither free fall more than 6 feet (1.8 meters) nor contact any lower level;
- Bring an employee to a complete stop and limit maximum deceleration distance an employee travels to 3.5 feet (1.07 meters); and
- Have sufficient strength to withstand twice the potential impact energy of an employee free falling a distance of 6 feet (1.8 meters) or the free fall distance permitted by the system, whichever is less.

The use of body belts for fall arrest is currently allowed, but effective January 1, 1998, the use of a body belt for fall arrest will be prohibited.

Personal fall arrest systems must be inspected prior to each use for wear damage, and other deterioration. Defective components must be removed from service. Dee-rings and snaphooks must have a minimum tensile strength of 5,000 pounds (22.2 kilonewtons). Dee-rings and snaphooks shall be proof-tested to a minimum tensile load of 3,600 pounds (16 kilonewtons) without cracking, breaking, or suffering permanent deformation.

Snaphooks shall be sized to be compatible with the member to which they will be connected, or shall be of a locking configuration.

Horizontal lifelines must be installed and used under the supervision of a qualified person. Self-retracting lifelines and lanyards that automatically limit free fall distance to 2 feet (0.61 meters) or less shall be capable of sustaining a minimim tensile load of 3,000 pounds (13.3 kilonewtons) applied to the device with the lifeline or lanyard in the fully extended position.

Self-retracting lifelines and lanyards that do not limit free fall distance to 2 feet (0.61 meters) or less, ripstitch lanyards, and tearing and deforming lanyards shall be capable of sustaining a minimum tensile load of 5,000 pounds (22.2 kilonewtons) applied to the device with the lifeline or lanyard in the fully extended position.

Anchorages shall be designed, installed, and used under the supervision of a qualified person, as part of a complete personal fall arrest system that maintains a safety factor of at least two, i.e., capable of supporting at least twice the weight expected to be imposed upon it. Anchorages used to attach personal fall arrest systems shall be independent of any anchorage being used to support or suspend platforms and must be capable of supporting at least 5,000 pounds (22.2 kilonewtons) per person attached.

Lanyards and vertical lifelines must have a minimum breaking strength of 5,000 pounds (22.2 kilonewtons).

Other acceptable fall protection systems under §1926.502 can be used. Such as Controlled Access Zones will be allowed as part of a fall protection plan when it is designed according to OSHA requirements. Safety monitoring systems will also be acceptable as part of a fall protection plan. Covers will be utilized to protect against fall hazards (like floor holes).

Written fall protection plans may be used when employers can demonstrate that conventional fall protection systems pose a greater hazard to employees than using alternative systems.

A warning line system can also be used. It consists of ropes, wires, or chains. Warning lines shall be erected around all sides of roof work areas. When mechanical equipment is being used, the warning line shall be erected not less than 6 feet (1.8 meters) from the roof edge parallel to the direction of mechanical equipment operation, and not less than 10 feet (3 meters) from the roof edge perpendicular to the direction of mechanical equipment operation.

When mechanical equipment is not being used, the warning line must be erected not less than 6 feet (1.8 meters) from the roof edge.

§1926.503 covers training requirements. Employers must provide a training program that teaches employees who might be exposed to fall hazards how to recognize such hazards and how to minimize them.

Employees must be trained in the following areas:

- the nature of fall hazards in the work area;
- the correct procedures for erecting, maintaining, disassembling, and inspecting fall protection systems;
- the use and operation of controlled access zones and guardrail, personal fall arrest, safety net, warning line, and safety monitoring systems;
- the role of each employee in the safety monitoring when the system is in use;
- the limitations on the use of mechanical equipment during the performance of roofing work on low-sloped roofs;
- the correct procedures for equipment and materials handling and storage and the erection of overhead protection; and
- employees' role in fall protection plans.

Employers must prepare a written certification that identifies the employee trained and the date of the training. The employer or trainer must sign the certification record. Retraining also must be provided when necessary.

13.   Subpart N. Cranes, Derricks, Hoists, Elevators, and Conveyors

SUBPART N - CRANES, DERRICKS, HOISTS, ELEVATORS, AND CONVEYORS	
1926.550	Cranes and derricks
1926.551	Helicopter cranes
1926.552	Material hoists, personnel hoists and elevators
1926.553	Base-mounted drum hoists
1926.554	Overhead hoists
1926.555	Conveyors
1926.556	Aerial lifts

Construction employers who use the kinds of lifting equipment and mechanisms identified in the titles of those 7 Subpart N standards must familiarize themselves with the applicable requirements and observe them.

§1926.556 applies to the following kinds of aerial devices that are used for elevating personnel to above-ground jobsites: (1) Extensible boom platforms, (2) Aerial ladders, (3) Articulating boom platforms, (4) Vertical towers, and (5) A combination of any of the first four.

14.   Subpart O. Motor Vehicles, Mechanized Equipment, and Marine Operations

SUBPART O - MOTOR VEHICLES, MECHANIZED EQUIPMENT, AND MARINE OPERATIONS	
1926.600	Equipment
1926.601	Motor vehicles
1926.602	Material handling equipment
1926.603	Pile driving equipment
1926.604	Site clearing
1926.605	Marine operations and equipment
1926.606	Definitions applicable to this subpart

The Subpart O standards apply to the equipment and operations listed in the titles of its 6 substantive standards.

§1926.600 imposes various requirements for (1) equipment that is left unattended at night, (2) when employees inflate, mount or dismount tires installed on split rims or rims equipped with locking rims or similar devices, (3) for heavy equipment or machinery that is parked, suspended or held aloft by slings, hoists or jacks, (4) for care and charging batteries, (5) when the equipment is being moved in the vicinity of power lines or energized transmitters and (6) when rolling cars are on spur railroad tracks. It also requires that cab glass must be safety glass (or equivalent) so that there will be no visible distortion to affect the safe operation of the machines that are regulated in Subpart O.

§1926.601 imposes various requirements for vehicles that operate within an off-highway jobsite that is not open to public traffic.

§1926.602 applies to earthmoving equipment, off-highway trucks, rollers, compactors, front-end loaders, bulldozers, tractors, lift-trucks, stackers, high-lift rider industrial trucks and similar construction equipment.

Pile driving and the equipment used in that operation is regulated by §1926.603.

§1926.604 requires rollover guards and canopy guards on equipment used in site clearing operations and provides that employees engaged in such operations must be protected from irritant and toxic plants and instructed in the first aid treatment that is available.

§1926.605 imposes various requirements upon employers engaged in marine construction operations.

OSHA rules for other vehicles frequently used in construction are included in Subpart W. See the discussion at page 213, infra.

15.   Subpart P. Excavations

SUBPART P - EXCAVATIONS	
1926.650	Scope, application, and definitions applicable to this subpart
1926.651	Specific excavation requirements
1926.652	Requirements for protective systems

The Subpart P standards apply to all open excavations (including trenches) made in the earth's surface. In some situations it requires the use of **written** designs (approved by a registered professional engineer) for sloping, benching and support systems.

There are 5 explanatory appendices to Subpart P that should be consulted by employers engaged in excavation work. They include illustrations, tables and diagrams. They cover the following subjects:

Appendix A: Soil Classification
Appendix B: Sloping and Benching
Appendix C: Timber Shoring for Trenches
Appendix D: Aluminum Hydraulic Shoring for Trenches
Appendix E: Alternatives to Timber Shoring
Appendix F: Selection of Protective Systems

The Subpart P standards went into effect in 1990. They constitute a total revision of the OSHA trenching and excavation standards that had been in effect for the previous 18 years.

§1926.651 has specific requirements for excavations that need to be reviewed for further details. Two things need to be remembered here. Daily inspections of excavations must be made by a competent person to make sure that cave-ins, failure of protective systems or hazardous atmospheres are not found.

If the competent person discovers there are indications that could result in a possible cave-in, exposed employees must be removed from the hazardous area until the necessary precautions have been taken.

16.   Subpart Q. Concrete and Masonry Construction

SUBPART Q - CONCRETE AND MASONRY CONSTRUCTION	
1926.700	Scope, application, and definitions applicable to this subpart
1926.701	General requirements
1926.702	Requirements for equipment and tools
1926.703	Requirements for cast-in-place concrete
1926.704	Requirements for precast concrete
1926.705	Requirements for lift-slab construction operations
1926.706	Requirements for masonry construction

Subpart Q includes detailed requirements to be observed in concrete and masonry construction operations.

A half-dozen general rules that apply to **all** such operations are set forth in §1926.701.

§1926.702 includes requirements for bulk storage bins, containers and silos, §1926.702(a); as well as Concrete Mixers, §1926.702(b); Powered and Rotating-Type Troweling Machines, §1926.702(c); Concrete Buggies, §1926.702(d); Concrete Pumping Systems, §1926.702(e); Concrete Buckets, §1926.702(f); Tremies, §1926.702(g); Bull Floats, §1926.702(h); Masonry Saws, §1926.702(i); and lockout/tagout procedures for compressors, mixers, and screens or pumps that are used in concrete and masonry construction activities, §1926.702(j).

The particular requirements for masonry construction are set forth in §1926.706.

Detailed requirements for three specialized types of operation (cast-in-place concrete, precast concrete, and lift-slab construction) are included in §1926.703 through §1926.705. Brief explanatory appendices are appended to both §1926.703 and §1926.705.

An appendix at the end of Subpart Q (Appendix A) lists 17 reference sources that can be helpful in understanding its various requirements.

17.   Subpart R. Steel Erection

SUBPART R - STEEL ERECTION	
1926.750	Flooring requirements
1926.751	Structural steel assembly
1926.752	Bolting, riveting, fitting-up, and plumbing-up

The Subpart R standards are rather brief and are limited in their application to steel erection operations. Employers so engaged should be familiar with those requirements.

8.   Subpart S. Underground Construction, Caissons, Cofferdams and Compressed Air

SUBPART S - UNDERGROUND CONSTRUCTION, CAISSONS, COFFERDAMS AND COMPRESSED AIR	
1926.800	Underground construction
1926.801	Caissons
1926.802	Cofferdams
1926.803	Compressed Air
1926.804	Definitions applicable to this subpart

Subpart S applies only to underground construction work including tunnels, shafts, chambers and passageways; caisson work; work on cofferdams; and work conducted in a compressed air environment. Decompression Tables are attached to Subpart S as Appendix A.

The four standards that make up Subpart S are very detailed so employers who do that kind of work should pay strict attention to them.

19. Subpart T. Demolition

SUBPART T - DEMOLITION	
1926.850	Preparatory operations
1926.851	Stairs, passageways, and ladders
1926.852	Chutes
1926.853	Removal of materials through floor openings
1926.854	Removal of walls, masonry sections, and chimneys
1926.855	Manual removal of floors
1926.856	Removal of walls, floors, and material with equipment
1926.857	Storage
1926.858	Removal of steel construction
1926.859	Mechanical demolition
1926.860	Selective demolition by explosives

The Subpart T standards are restricted in their application to employers engaged in demolition operations. They are neither lengthy nor complicated. Employers engaged in those operations should have little difficulty understanding them. However, if they do, there is free consultation available in every state that is as close as your telephone. The address and telephone numbers for each state are listed in part three of this book.

20. **Subpart U. Blasting and Use of Explosives**

SUBPART U - BLASTING AND USE OF EXPLOSIVES	
1926.900	General provisions
1926.901	Blaster qualifications
1926.902	Surface transportation of explosives
1926.903	Underground transportation of explosives
1926.904	Storage of explosives and blasting agents
1926.905	Loading of explosives or blasting agents
1926.906	Initiation of explosive charges - electrical blasting
1926.907	Use of safety fuse
1926.908	Use of detonating cord
1926.909	Firing the blast
1926.910	Inspection after blasting
1926.911	Misfires
1926.912	Underwater blasting
1926.913	Blasting in excavation work under compressed air
1926.914	Definitions applicable to this subpart

Subpart U applies when employers use explosives or do blasting in their work. No employer should use explosives unless he first familiarizes himself with the Subpart U standards. They impose detailed restrictions on those working with explosives.

OSHA has identified parts of 4 "general industry" standards that apply to Subpart U:
- Buildings used for blasting agent mixing, §1910.109(g)(2)(ii),
- Buildings used for the mixing of water gels, §1910.109(h)(3)(ii),
- Semiconductive hose for explosives and blasting agents, §1910.109(a)(12), and
- Pneumatic blasting agent loading over blasting caps, §1910.109(e)(3)(iii).

21.   Subpart V. Power Transmission and Distribution

SUBPART V - POWER TRANSMISSION AND DISTRIBUTION	
1926.950	General Requirements
1926.951	Tools and protective equipment
1926.952	Mechanical equipment
1926.953	Material handling
1926.954	Grounding for protection of employees
1926.955	Overhead lines
1926.956	Underground lines
1926.957	Construction in energized substations
1926.958	External load helicopters
1926.959	Lineman's body belts, safety straps, and lanyards
1926.960	Definitions applicable to this subpart

The Subpart V standards are limited in their application to construction, erection, alteration, conversion and improvement of electric transmission and distribution lines and equipment. Employers engaged in that type of work must familiarize themselves with those standards. Employers who do not do work of that kind will have no reason to review the Subpart V standards.

22.   Subpart W. Rollover Protective Structures; Overhead Protection

SUBPART W - ROLLOVER PROTECTIVE STRUCTURES; OVERHEAD PROTECTION	
1926.1000	Rollover protective structures (ROPS) for material handling equipment
1926.1001	Minimum performance criteria for rollover protective structures for designated scrapers, loaders, dozers, graders, and crawler tractors
1926.1002	Protective frame (ROPS) test procedures and performance requirements for wheel-type agricultural and industrial tractors used in construction
1926.1003	Overhead protection for operators of agricultural and industrial tractors

Subpart W only applies to seven kinds of material handling equipment when used in construction work: (1) rubber-tired, self-propelled scrapers; (2) rubber-tired front-end loaders; (3) rubber-tired dozers; (4) wheel-type agricultural and industrial tractors; (5) crawler tractors; (6) crawler-type loaders; and (7) motor graders, with or without attachments.

Sideboom pipelaying tractors are **not** regulated by Subpart W.

Employers who use the seven kinds of equipment listed above should familiarize themselves with the applicable requirements. Those who don't use that equipment should have no reason to be concerned with Subpart W.

23.   Subpart X. Stairways and Ladders

SUBPART X - STAIRWAYS AND LADDERS	
1926.1050	Scope, application, and definitions applicable to this subpart OSHA has identified the "general industry" standard regulating Compressed Air Receivers, §1910.169, and the standard that requires bumper (or derail) blocks on spur railroad tracks, §1910.176(f), as applicable to Subpart O.
1926.1051	General requirements
1926.1052	Stairways
1926.1053	Ladders
1926.1060	Training requirements

The Subpart X standards went into effect in 1991. They set forth the conditions and circumstances under which ladders and stairways must be provided (and used). The standards apply to all stairways and ladders used in construction, alteration, repair, demolition, painting and repair work (however, some of the requirements do not apply to some ladders and stairways that were built prior to March 15, 1991).

Any employer who has one or more employees engaged in construction work who will use a ladder or stairway while working -- and that covers the vast majority of construction employers -- must become familiar with the Subpart X requirements.

§1926.1060 requires that a **training** program be provided for each employee using ladders and stairways so that he will be able to recognize the hazards related to ladders and stairways and know the procedures to be followed in order to minimize those hazards.

An appendix (Appendix A) has been added to Subpart X to assist employers in complying with the loading and strength requirements for ladders that are set forth in §1926.1053(a)(1).

24.   Subpart Y. Diving

Subpart Y is identical to the Part 1910 "General Industry" standards discussed earlier in

this book under the heading: "Subpart T. Commercial Diving Operations."

25.   Subpart Z. Toxic and Hazardous Substances

Subpart Z contains 24 substance-specific standards that are identical to the "General

Industry" standards that regualte those same substances (see the earlier discussion of Part 1910,

Subpart Z, Toxic and Hazardous Substances). Those standards are designated as follows in the

construction industry standards.

§1926.1101	Asbestos
§1926.1102	Coal tar pitch volatiles: interpretation of term.
§1926.1103	4-Nitrobiophenyl.
§1926.1104	alpha-Naphthylamine.
§1926.1106	Methyl chloromethyl ether.
§1926.1107	3.3'-Dichlorobenzidine (and its salts).
§1926.1108	bis-Chloromethyl ether.
§1926.1109	beta-Naphthylamine.
§1926.1110	Benzidine.
§1926.1111	4-Aminodiphenyl.
§1926.1112	Ethyleneimine.
§1926.1113	beta-Propiolactone.
§1926.1114	2-Acetylaminofluorene.
§1926.1115	4-Dimethylaminoazobenzene.
§1926.1116	N-Nitrosodimethylamine.
§1926.1117	Vinyl chloride.
§1926.1118	Inorganic arsenic.
§1926.1127	Cadmium
§1926.1128	Benzene.
§1926.1129	Coke oven emissions.
§1926.1144	1,2 dibromo-3-chloropropane.
§1926.1145	Acrylonitrile.
§1926.1147	Ethylene oxide.
§1926.1148	Formaldehyde.

VI.   OSHA AGRICULTURE STANDARDS

There are relatively few occupational safety and health standards that are applicable to agricultural operations. They are included in Part 1928 of Title 29, Code of Federal Regulations..

It has only 5 Subparts as follows:

- Subpart A -   General
- Subpart B -   Applicability of Standards
- Subpart C -   Roll-Over Protective Structures
- Subpart D -   Safety for Agricultural Equipment
- Subpart I* -   General Environmental Controls

Subpart A does not contain any requirements.

Subpart B provides that the following five "general industry" standards apply to agricultural operations:

1. Temporary labor camps - §1910.142
2. Storage and handling of anhydrous ammonia - §1910.111(a) and (b)
3. Pulpwood logging - §1910.266
4. Slow-moving vehicles - §1910.145
5. Hazard communication - §1910.1200

Subpart C contains 1 standard, §1928.51. It requires that agricultural tractors manufactured after October 25, 1976 must be equipped with roll-over protective structures (ROPS) and seatbelts, and must have their batteries, fuel tanks, oil reservoirs and coolant systems so constructed and located (or sealed) to assure against spillage that might come in contact with the tractor operator if an upset occurs. The edges and corners at the operator's station must also be designed to minimize operator injury in the event of an upset. Certain "low profile" tractors do not have to have ROPS or seatbelts when used in some orchards, vineyards, hop yards, farm buildings and greenhouses.

Subpart C also includes an appendix (Appendix A) that contains operating instructions for employees who drive tractors, and another appendix (Appendix B) with 16 illustrations designed to aid in understanding the Subpart C requirements.

Subpart D consists of a single standard, §1928.57, that contains the guarding, maintenance and servicing requirements for moving machinery parts of farm field equipment, farmstead equipment, and cotton gins used in agricultural operations. It also requires that employers provide instruction in safe operating practices to employees who operate and service that equipment.

Subpart I contains a single standard, §1928.110, that imposes field sanitation requirements on any agricultural establishment where, on any given day, there are eleven or more employees engaged in hand-labor operations in the field. The field sanitation requirements include potable drinking water as well as toilet and washing facilities that are maintained in accordance with appropriate public health sanitation practices. Employers are also required to inform their employees of good hygiene practices, advise them of the location of the water and sanitation facilities and allow the employees "reasonable opportunities" to use them during the workday.

Subpart M contains one standard §1928.1027. Cadmium imposes many detailed requirements. If your company uses this, it is best to look into this regulation and find out what some of the details are.

PART THREE

STATE BY STATE OSHA DIRECTORY

```
                  NOTE

Prior to seeking the assistance of any
OSHA Consultant, you should read the
discussion entitled "Consultative
Services" in Part One of this book.
```

Contacting OSHA Offices

LOCAL AREA OFFICES

The following is a list of addresses and telephone numbers of OSHA Area Offices and Consultation Offices. These offices are sources of information, publications, and assistance in understanding the requirements of the standards.

*Each area OSHA office is also a source of information, publications and assistance in understanding OSHA requirements. It can furnish you with the basic publications you need, such as: 1. The OSHA workplace poster: "Job Safety and Health Protection." 2. The OSHA recordkeeping requirements. 3. Copies of OSHA standards. 4. A large selection of publications concerned with safe work practices, control of hazardous substances, employer and employee rights and responsibilities, and other subjects.

You can feel free to contact the OSHA Area Office by telephone, by mail or in person, without fear of triggering an OSHA inspection. However, do not request OSHA to visit your workplace. Such a visit would require citations and fines if a violation existed. If you want advice or on-site consultation, contact the **CONSULTATION** service for your state.

ALABAMA

ENFORCEMENT OFFICE
OSHA Area Office*
U.S. Department of Labor
2047 Canyon Road - Todd Mall
Birmingham, Alabama 35216
Telephone: (205) 731-1534

CONSULTATION OFFICE
Onsite Consultation Program
425 Martha Parham West
Tuscaloosa, Alabama 35487
Telephone: (205) 348-3033

ALASKA

ENFORCEMENT OFFICE
Alaska Department of Labor
1111 West 8th Street
Room 306
Juneau, Alaska 99801
Telephone: (907) 465-2700

CONSULTATION OFFICE
Division of Consultation and Training
3301 Eagle Street
Anchorage, Alaska 99510
Telephone: (907) 264-2599

ARIZONA

ENFORCEMENT OFFICE
Division of Occupational Safety & Health
Industrial Commission of Arizona
800 W. Washington Street
Phoenix, Arizona 85007
Telephone: (602) 542-5795

CONSULTATION OFFICE
Division of Consultation and Training
Industrial Commission of Arizona
800 W. Washington Street
Phoenix, Arizona 85007

ARKANSAS

ENFORCEMENT OFFICE
OSHA Area Office*
U.S. Department of Labor
425 W. Capitol, Suite 450
Little Rock, Arkansas 72201
Telephone: (501) 324-6291

CONSULTATION OFFICE
OSHA Consultation
Arkansas Department of Labor
10421 West Markham
Little Rock, Arkansas 72205
Telephone: (501) 682-4522

CALIFORNIA

ENFORCEMENT OFFICE
California Department of Industrial
Relations
71 Stevenson Street, Room 415
San Francisco, California 94105
(415) 744-7120

CONSULTATION OFFICE
CAL/OSHA Consultation Service
455 Golden Gate Avenue
4th Floor
San Francisco, California 94102
(415) 703-4590

COLORADO

ENFORCEMENT OFFICE
OSHA Area Office*
U.S. Department of Labor
1391 N. Speer Blvd. - Suite 210
Denver, Colorado 80204
Telephone: (303) 844-5285

CONSULTATION OFFICE
Occupational Safety and Health Section
Colorado State University
110 Veterinary Science Bldg.
Fort Collins, Colorado 80523
Telephone: (303) 491-6151

CONNECTICUT

ENFORCEMENT OFFICE
OSHA Area Office*
U.S. Department of Labor
One Lafayette Square, Suite 202
Bridgeport, Connecticut 06604
Telephone: (203) 579-5579

CONSULTATION OFFICE
Division of Occupational Safety and Health
Consultation Services
200 Folly Brook Blvd.
Wethersfield, Connecticut 06109
Telephone: (203) 566-4550

DELAWARE

ENFORCEMENT OFFICE
OSHA Area Office*
U.S. Department of Labor
1 Rooney Square, Suite 402
920 King Street
Wilmington, Delaware 19801
Telephone: (302) 573-6115

CONSULTATION OFFICE
Occupational Safety and Health
Consultation Program
Delaware Department of Labor
820 North French Street, 6th Floor
Wilmington, Delaware 19801
Telephone: (302) 577-3908

FLORIDA

ENFORCEMENT OFFICE
OSHA Area Office*
U.S. Department of Labor
8040 Peters Road
Fort Lauderdale, Florida 33324
Telephone: (305) 424-0242

250

FLORIDA - Continued

CONSULTATION OFFICE
On-site Consultation Program
Department of Labor & Employment
2002 St. Augustine Road
Building E, Suite 45
Tallahassee, Florida 32399
Telephone: (904) 488-3044

GEORGIA

ENFORCEMENT OFFICE
OSHA Area Office*
U.S. Department of Labor
La Vista Perimeter Office Park
Bldg. 7, Suite 110
Tucker, Georgia 30084
Telephone: (404) 493-6644

CONSULTATION OFFICE
On-Site Consultation Program
Georgia Institute of Technology
O'Keefe Building, Room 23
Atlanta, Georgia 30332
(404) 894-8274

GUAM

ENFORCEMENT OFFICE
OSHA Area Office*
U.S. Department of Labor
300 Ala Moana Blvd., Suite 5122
Honolulu, Hawaii 96850
Telephone: (808) 541-2685

CONSULTATION OFFICE
OSHA On-Site Consultation
Guam Departmen tof Labor
3rd Floor ITC Building
Tamuning, Guam 96911
Telephone: (671) 646-9244

HAWAII

ENFORCEMENT OFFICE
Hawaii Department of Labor
300 Ala Moana Blvd., Suite 5122
Honolulu, Hawaii 96850
Telephone: (808) 541-3456

CONSULTATION OFFICE
Consultation and Training Branch
Hawaii Department of Labor and
Industrial Relations
830 Punchbowl Street
Honolulu, Hawaii 96813
Telephone: (808) 586-8844

IDAHO

ENFORCEMENT OFFICE
OSHA Area Office*
U.S. Department of Labor
3050 North Lake Harbor Lane
Suite 134
Boise, Idaho 83903
Telephone: (208) 334-1867

CONSULTATION OFFICE
Safety & Health Consultation Project
Boise State University
Department of Community & Environment
Health
1910 University Drive, M6-110
Boise, Idaho 83725
Telephone: (208) 385-3283

ILLINOIS

ENFORCEMENT OFFICE
OSHA Area Office*
U.S. Department of Labor
2360 E. Devon Avenue
Suite 1010
Des Plaines, Illinois 60018
Telephone: (312) 803-4800

251

ILLINOIS - Continued

CONSULTATION OFFICE
Illinois On-Site Consultation
Industrial Services Division
Department of Commerce & Community
Affairs
State of Illinois Center
100 West Randolph Street, Suite 3-400
Chicago, Illinois 60601
Telephone: (312) 814-2337

INDIANA

ENFORCEMENT OFFICE
Indiana Department of Labor
46 E. Ohio Street
Room 423
Indianapolis, Indiana 46204
Telephone: (317) 226-7290

CONSULTATION OFFICE
Indiana Department of Labor
Bureau of Safety, Education & Training
402 West Washington Street
Room W195
Indianapolis, Indiana 46204
Telephone: (317) 232-2378

IOWA

ENFORCEMENT OFFICE
Iowa Department of Labor
210 Walnut Street, Room 815
Des Moines, Iowa 50309
Telephone: (515) 284-4794

CONSULTATION OFFICE
Iowa Bureau of Labor
Consultation Program
1000 E. Grand Avenue
Des Moines, Iowa 50319
Telephone: (515) 281-3447

KANSAS

ENFORCEMENT OFFICE
OSHA Area Office*
U.S. Department of Labor
8600 Farley, Suite 105
Overland Park, Kansas 66212
Telephone: (913) 236-3220

CONSULTATION OFFICE

Kansas 7(c)(1) Consultation Program
Kansas Department of Human Resources
512 West 6th Street
Topeka, Kansas 66603
Telephone: (913) 296-4386

KENTUCKY

ENFORCEMENT OFFICE
Kentucky Division of Labor
John C. Watts Federal Building
330 W. Broadway - Room 108
Frankfort, Kentucky 40601
Telephone: (502) 227-7024

CONSULTATION OFFICE
Kentucky Labor Cabinet
1049 U.S. Highway 127 South
Frankfort, Kentucky 40601
Telephone: (502) 564-3070

LOUISIANA

ENFORCEMENT OFFICE
OSHA Area Office*
U.S. Department of Labor
2156 Wooddale Boulevard, Suite 200
Baton Rouge, Louisiana 70806
Telephone: (504) 389-0463

CONSULTATION OFFICE
Consultation Program
Lousiana Dept. of Employment & Training
P.O. Box 94094
Baton Rouge, Louisiana 70804
Telephone: (504) 342-9601

MAINE

ENFORCEMENT OFFICE
OSHA Area Office*
U.S. Department of Labor
202 Harlow Street
Room 211
Bangor, Maine 04401
Telephone: (207) 941-8177

CONSULTATION OFFICE
Consultation Program
Division of Industrial Safety
Maine Department of Labor
State Home Station 82
Hallowell Annex
Augusta, Maine 04333
Telephone: (207) 289-6460

MARYLAND

ENFORCEMENT OFFICE
Maryland Occupational Safety & Health
31 Hopkins Plaza
Federal Building, Room 1110
Baltimore, Maryland 21201
Telephone: (410) 962-2840

CONSULTATION OFFICE
Maryland Division of Labor
501 St. Paul Place - 2nd Floor
Baltimore, Maryland 21202
Telephone: (410) 333-4179

MASSACHUSETTS

ENFORCEMENT OFFICE
OSHA Area Office*
U.S. Department of Labor
1145 Main Street
Room 108
Springfield, Massachusetts 01103
Telephone: (413) 785-0123

CONSULTATION OFFICE
Consultation Program
Division of Industrial Safety
MA Department of Labor Industries
100 Cambridge Street
Boston, Massachusetts 02202
Telephone: (617) 727-3463

MICHIGAN

ENFORCEMENT OFFICE
MIOSHA
801 S. Waverly Road
Suite 306
Lansing, Michigan 48917
Telephone: (517) 377-1892

CONSULTATION OFFICE
Michigan Dept. of Labor
Victor Office Center
201 N. Washington Square
P.O. Box 30015
Lansing, Michigan 48933
Telephone: (517) 373-9600

MINNESOTA

ENFORCEMENT OFFICE
Minnesota OSH Division
110 South Fourth Street
Room 116
Minneapolis, Minnesota 55401
Telephone: (612) 348-1994

CONSULTATION OFFICE
Minnesota Department of Labor
443 Lafayette Road
St. Paul, Minnesota 55155
Telephone: (612) 296-2342

MISSISSIPPI

ENFORCEMENT OFFICE
OSHA Area Office*
U.S. Department of Labor
3780 I-55 North
Room 210
Jackson, Mississippi 39211
Telephone: (601) 965-4606

CONSULTATION OFFICE
On-Site Consultation Program
Division of Occupational Safety & Health
Mississippi State Board of Health
2906 N. State Street, Suite 201
Jackson, Mississippi 39216
Telephone: (601) 987-3981

MISSOURI

ENFORCEMENT OFFICE
OSHA Area Office*
U.S. Department of Labor
6200 Connecticut, Room 100
Kansas City, Missouria 64120
Telephone: (816) 483-9531

CONSULTATION OFFICE
On-Site Consultation Program
Division of Labor & Standards
Dept. of Labor and Industrial Relations
3315 West Truman Boulevard
Jefferson City, Missouri 65109
Telephone: (314) 751-3403

MONTANA

ENFORCEMENT OFFICE
OSHA Area Office*
U.S. Department of Labor
19 N. 25th Street
Billings, Montana 59101
Telephone: (406) 657-6389

CONSULTATION OFFICE
Consultation Program
Division of Labor and Industry
Employment Relations Division
Safety Bureau
Arcade Building, 111 North Main
Helena, Montana 59604
Telephone: (406) 444-6401

NEBRASKA

ENFORCEMENT OFFICE
OSHA Area Office*
U.S. Department of Labor
Overland-Wolf Building - Room 100
6910 Pacific Street
Omaha, Nebraska 68106
Telephone: (402) 221-3182

CONSULTATION OFFICE
Consultation Services
Div. of Safety Labor & Safety Standards
State Office Building, Lower Level
301 Centennial Mall, South
Lincoln, Nebraska 68509
Telephone: (402) 471-4717

NEVADA

ENFORCEMENT OFFICE
Div. of Enforcement for Industrial
Safety and Health
4600 Kietzke Lane, Building F
Suite 153
Reno, Nevada 89502
Telephone: (702) 688-1380

CONSULTATION OFFICE
Division of Preventive Safety
Dept of Industrial Relations
2500 W. Washington Street, Suite 106
Las Vegas, Nevada 89106
Telephone: (702) 486-5066

NEW HAMPSHIRE

ENFORCEMENT OFFICE
OSHA Area Office*
U.S. Department of Labor
279 Pleasant Street, Room 201
Concord, New Hampshire 03301
Telephone: (603) 225-1629

CONSULTATION OFFICE
On-Site Consultation Program
New Hampshire Department of Labor
State Office Park South
95 Pleasant Street
Concord, New Hampshire 03301
Telephone: (603) 271-2024

NEW JERSEY

ENFORCEMENT OFFICE
OSHA Area Office*
U.S. Department of Labor
Plaza 35, Suite 205
1030 Saint Georges Avenue
Avenel, New Jersey 07001
Telephone: (201) 750-3270

CONSULTATION OFFICE
Consultation Services
Division of Workplace Standards
New Jersey Department of Labor
CN 386
Trenton, New Jersey 08625

NEW MEXICO

ENFORCEMENT OFFICE
Occupational Health & Safety Bureau
505 Marquette Avenue, N.W.
ALbuquerque, New Mexico 87102
Telephone: (505) 766-3411

CONSULTATION OFFICE
Occupational Health & Safety Bureau
1190 St. Francis Drive, Room N2200
Santa Fe, New Mexico 87503
Telephone: (505) 827-2877

NEW YORK

ENFORCEMENT OFFICE
OSHA Area Office*
U.S. Department of Labor
90 Church Street, Room 1407
New York, N.Y. 10007
Telephone: (212) 264-9840

CONSULTATION OFFICE
Consultation Services
Division of Safety and Health
State Office Campus
Building 12, Room 457
Albany, New York 12240
Telephone: (518) 457-3518

NORTH CAROLINA

ENFORCEMENT OFFICE
N.C. Division of Occupational
Safety and Health
300 Fayetteville Street Mall, Room 438
Raleigh, North Carolina 27601
Telephone: (919) 856-4770

CONSULTATION OFFICE
N.C. Department of Labor
319 Chapanoke Road
Raleigh, North Carolina 27603
Telephone: (919) 662-4585

NORTH DAKOTA

ENFORCEMENT OFFICE
OSHA Area Office*
U.S. Department of Labor
220 E. Rosser, P.O. Box 2439
Bismarck, North Dakota 58502
Telephone: (701) 250-4521

CONSULTATION OFFICE
Consultation Services
N.D. State Dept. of Health
1200 Missouri Avenue, Room 304
Bismarck, North Dakota 58502
Telephone: (701) 221-5188

255

OHIO

ENFORCEMENT OFFICE
OSHA Area Office*
U.S. Department of Labor
Federal Office Building, Room 899
1240 East Ninth Street
Cleveland, Ohio 44199
Telephone: (216) 522-3818

CONSULTATION OFFICE
Division of On-Site Consultation
Department of Industrial Relations
2323 West Fifth Avenue
Columbus, Ohio 43216
Telephone: (614) 644-2631

OKLAHOMA

ENFORCEMENT OFFICE
OSHA Area Office*
U.S. Department of Labor
420 West Main Place
Suite 300
Oklahoma City, Oklahoma 73102
Telephone: (405) 231-5351

CONSULTATION OFFICE
Consultation Services, OSHA Division
Oklahoma Department of Labor
4001 North Lincoln Boulevard
Oklahoma City, Oklahoma 73105
Telephone: (405) 528-1500

OREGON

ENFORCEMENT OFFICE
OR-OSHA
9500 S.W. Barbur Blvd., Suite 200
Portland, Oregon 97219
Telephone: (503) 229-5910

CONSULTATION OFFICE
OR-OSHA, Dept. of Insurance & Finance
21 Labor and Industries Bldg.
Salem, Oregon 97310
Telephone: (503) 378-3272

PENNSYLVANIA

ENFORCEMENT OFFICE
OSHA Area Office*
U.S. Department of Labor
Progress Plaza
49 N. Progress Avenue
Harrisburg, Pennsylvania 17109
Telephone: (717) 782-3902

CONSULTATION OFFICE
PA/OSHA Consultation Program
Idiana Univ. of Pennsylvania
205 Uhler Hall
Indiana, Pennsylvania 15705
Telephone: (215) 775-9704

PUERTO RICO

ENFORCEMENT OFFICE
Puerto Rico Department of Labor
U.S. Courthouse Federal Office Building
Carlos Chandon Ave, Room 555
Hato Rey, Puerto Rico 00918
Telephone: (809) 766-5457

CONSULTATION OFFICE
Puerto Rico OSH Office
Voluntary Programs Division
505 Munoz Riveria Avenue
Hato Rey, Puerto Rico 00918
Telephone: (809) 754-2119

RHODE ISLAND

ENFORCEMENT OFFICE
OSHA Area Office*
U.S. Department of Labor
380 Westminister Mall, Room 243
Providence, Rhode Island 02903
Telephone: (401) 528-4669

CONSULTATION OFFICE
Consultation Services, RI Dept. of Health
206 Cannon Bldg., 75 Davis Street
Providence, Rhode Island 02908
Telephone: (401) 277-2438

SOUTH CAROLINA

ENFORCEMENT OFFICE
South Carolina Department of Labor
1835 Assembly Street, Room 1468
Columbia, South Carolina 29201
Telephone: (803) 765-5904

CONSULTATION OFFICE
On-Site Consultation Program
South Carolina Department of Labor
3600 Forest Drive
Columbia, South Carolina 29211
Telephone: (803) 734-9599

SOUTH DAKOTA

ENFORCEMENT OFFICE
OSHA Area Office*
U.S. Department of Labor
19 North 25th Street
Billings, Montana 59101
Telephone: (406) 657-6649

CONSULTATION OFFICE
Consultation Services
S.T.A.T.E. Engineering Extension
On-Site Technical Division
South Dakota State University
Brookings, South Dakota 57007
Telephone: (605) 688-4101

TENNESSEE

ENFORCEMENT OFFICE
T-OSHA
710 James Robertson Pkwy, Gateway Plaza
Suite "A" , 2nd Floor
Nashville, Tennessee 37243

CONSULTATION OFFICE
OSHA Consultative Services
T-OSHA
501 Union Bldg., 6th Floor
Nashville, Tennessee 37219
Telephone: (615) 741-7036

TEXAS

ENFORCEMENT OFFICE
OSHA Area Office*
U.S. Department of Labor
611 East 6th Street, Room 303
Austin, Texas 78701

CONSULTATION OFFICE
Consultation Services
Texas Workers' Comp. Comm.
Health & Safety Division
Smithfield Building
4606 South IH 35
Austin, Texas 78704
Telephone: (512) 440-3834

UTAH

ENFORCEMENT OFFICE
Utah-OSHA
1781 S. 300 West
Salt Lake City, Utah 84165
Telephone: (801) 487-0267

CONSULTATION OFFICE
Utah Safety & Health Consul. Service
160 East 300 South, 3rd Floor
Salt Lake City, Utah 84114
Telephone: (801) 530-6868

VERMONT

ENFORCEMENT OFFICE
V-OSHA
Department of Labor & Industry
National Life Bldg., Drawer 20
Montpelier, VT 05620
Telephone: (802) 828-2765

CONSULTATION OFFICE
Consultation Services
Division of Occupational Safety & Health
Vermont Dept. of Labor & Industry
118 State Street
Montpelier, Vermont 05602
Telephone: (802) 828-2765

257

VIRGINIA

ENFORCEMENT OFFICE
V-OSHA Field Office
Powers-Taylor Building
13 South Thirteenth Street
Richmond, Virginia 23219
Telephone: (804) 371-2327

CONSULTATION OFFICE
VA Department of Labor & Industry
Voluntary Safety & Health Compliance
13 S. 13 Street
Richmond, Virginia 23219
Telephone: (804) 786-8707

VIRGIN ISLANDS

ENFORCEMENT OFFICE
Occupational Safety & Health Division
Department of Labor
Government Complex, Bldg. 2, 2nd Floor
Lagoon Street, Room 207
Frederiksted, St. Croix, VI 00840
Telephone: (804) 772-1315

CONSULTATION OFFICE
Consultation Services
Division of Occupational Safety & Health
Virgin Islands Dept. of Labor
2131 Hospital Street, Box 890
Christian Street
St. Croix, Virgin Islands 00840
Telephone: (809) 773-1994

WASHINGTON

ENFORCEMNET OFFICE
Seattle (Region 2), WISHA Office
300 West Harrison
Seattle, Washington 98119
Telephone: (206) 281-5470

CONSULTATION OFFICE
Voluntary Services
Washington Dept. of Labor & Industries
1011 Plum Street, M/S HC-462
Olympia, Washington 98504
Telephone: (206) 586-0961

WEST VIRGINIA

ENFORCEMENT OFFICE
OSHA Area Office*
U.S. Department of Labor
550 Eagan Street, Room 206
Charleston, WV 25301
Telephone: (304) 347-3797

CONSULTATION OFFICE'
Consultation Services
WV Dept. of Labor
State Capitol Bldg., 3, Room 319
1800 E. Washington Street
Charleston, WV 25305
Telephone: (304) 558-7890

WISCONSIN

ENFORCEMENT OFFICE
OSHA Area Office*
U.S. Department of Labor
310 West WIsconsin Avenue
Suite 1180
Milwaukee, WI 53203
Telephone: (414) 297-3315

CONSULTATION OFFICE
Occupational Health Consultive Services
WI Dept. of Health & Human Services
Section of Occupational Health
1414 E. Washington Avenue
Room 112
Madison, Wisconsin 53703
Telephone: (608) 266-8579

WYOMING

ENFORCEMENT OFFICE
Wyoming Occupational Health &
Safety Department
Herschler Building 2 East
122 West 25th Street
Cheyenne, Wyoming 82002
Telephone: (307) 777-7786

CONSULTATION OFFICE
Department of Employment
122 West 25th Street
Herschler Bldg.
2nd Floor, East Wing
Cheyenne, Wyoming 82002
Telephone: (307) 777-7672

PART FOUR

SAFETY ON THE INTERNET

INTRODUCTION

The information superhighway is not a single entity, but an interconnected web of communications networks, computers, and databases that can put vast amounts of information at people's fingertips. Over time, access to these resources will change the way safety and health professionals work on a daily basis.

Safety Goes Online

The Internet is an exciting place for safety and health professionals. Each day brings new information, more Web sites and more people who go online with something to contribute. Since the advent of sophisticated Web-browser software a few years ago, what was once the domain of academics and government researchers has become an essential resource for anyone in a safety-related field.

The question is no longer should you get on the Net, but how do you handle the information overload? In a field where nobody has enough time, how do you decide when and where to surf?

Internet Support Services

The Internet is a worldwide network of interconnected computer networks and is considered by many to be the prototype of the future information superhighway. It is comprised of thousands of separately administered networks of many sizes and types; the total number of Internet users is in the millions and, according to one estimate, is growing

264

at the rate of one million users a month. The Internet currently supports three basic types

of services:

- Electronic mail (E-mail) - Electronic messages can be sent to any individual Internet user or can be broadcast to thousands of users simultaneously.

- Remote log-in (or telnet) - This is the ability to connect to and use a computer at a remote site. Once connected, the user can run programs or access information on that remote computer.

- File transfer protocol (or ftp) - This allows users to download, or transfer, data from a remote computer to a local computer. This includes shareware programs, documents, electronic journals, graphic images, and catalogs.

I.   WHAT IS THE INTERNET?

Is it surfing the waves on the beaches of Hawaii? We wish. In short, it's a system

of computer networks linked together from all over the world. That means if your

computer is connected to the Internet, you can view the files at thousands of other

Internet sites that are set up to make information publicly available. You can connect to

these computers in a matter of seconds, whether they are around the corner or on the

other side of the world.

1.   Accessing the Internet

To use the Internet, you need to have access to the Internet. This can be

accomplished in many ways - whether by a company connection or by opening an

To use the Internet, you need to have access to the Internet. This can be accomplished in many ways - whether by a company connection or by opening an account with one of the major online services such as *America Online*, *Compuserve* or *Prodigy*. The type of connection you have, as well as hardware considerations such as modern speed, will determine what Internet services you can use.

2. Mastering Internet Skills

Once you have Internet access, you need to master the skills necessary to use the Internet. This means learning how to use the various resources that are out there. People often ask, how much do I really need to learn? You only need to learn how to use the resources in which you are interested. You will also likely become interested in many of the resources that are available once you see what's out there.

3. Is it Hard to Learn How to Use the Internet?

No, it is not hard at all. It just takes practice. Don't be put off by people who say you have to have a degree in computer science or the Internet is not "user friendly." The Internet is one of the most important and complex inventions of mankind. Using the Internet really means learning some basic concepts and then teaching yourself how to use a variety of different programs.

can make a complex system so easy to learn that you can use it on the first day, is by removing (or hiding) most of the power. But then, once you become experienced, you find that the system is too simple and awkward.

Millions of people around the world already use the Internet. You don't need to be a computer expert. So to put this in perspective, using the Internet is a lot easier than many things that we all do every day, such as driving a car or shopping for groceries. All you need is some practice and some patience.

Our best advice? Open this book and find a Web site that interests you. Then start *surfing the net*. Experiment. Have fun. Go slow. Enjoy!

II. GETTING STARTED

Most computer applications have a foundation in familiar, everyday tasks. A word processor has the familiar feel of a typewriter. Money management software is based on a computer version of our friendly, if sometimes overtaxed, checkbook. It is easy to use these programs on a basic level. But before we perform the familiar task of placing our electronic letter in the mail, we are confronted with the technical details of computer communications. Computer telecommunications does **not** have to be intimidating. Let's cover the basics to help you go "on-line."

1. Modems

A modem and telecommunications software are your computer's doorway to cyberspace. A modem connects your computer to the telephone line, allowing you to exchange information with any other computer connected to a telephone line. Telecommunications software coordinates the information transfer between the computers.

Modems and their prices vary principally on how fast the modem can send and receive information, and on whether the modem can send and receive faxes. Raw speed is important because the major cost of computer communications is the cost of long-distance telephone calls to access other computers and any "connect" charges for the time you are connected to the remote computer. Presently, the maximum speed of most computer bulletin boards is 9,600 to 28,800 baud.

You should buy the fastest modem you can afford. Since communicating at 28,800 baud is 4 times faster than 9,600 baud, a high-speed modem can quickly pay for itself in telephone and connect charges. When buying a modem be careful to read the fine print. Some fax modems handle faxes at 9,600 baud but data at only 4,800 baud. It is the data baud rate that saves you money.

Basic telecommunications software is packaged with most modems. This software, or the terminal program in *Microsoft Windows*, is all you need to begin; but experienced users will want more powerful software such as *Procomm* or *Qmodem*.

2.   Communications Protocol

After the modem is connected to the computer and telephone line and the software installed on your computer, you must confront the most intimidating parts of computer communications: communication protocols. Fortunately, you don't need a technical understanding of the protocols. Just follow the instructions for your software to set it to the settings required by the computer you are calling.

Communications protocols determine how the modem communicates over the telephone lines, how information is displayed on your computer screen, and how computer files are transferred between computers. The two computers must agree on these protocols for successful communications.

Modem settings include the modem's speed, or baud rate; the number of data bits; whether there is a stop rate; and the type of parity. Most systems require 8 data bits, no parity, and 1 stop bit, commonly abbreviated as 8-N-1. Set your software to communicate at your modem's top speed. If a slower speed modem answers your call, your modem will automatically adjust to the lower speed.

How information is displayed on your computer screen, or terminal, is the terminal emulation protocol. The most common terminal emulations are TTY, ANSI, and VT100. I generally set my software to ANSI and change it to TTY or VT100 if the screen looks all mixed up. Make sure the software is set to wrap long lines.

Finally, there are protocols for transferring programs and computer files between computers. Simple text files can be transferred as text, but word processing files and computer programs must be transferred using special protocols. These special file transfer protocols also check the received informaiton for errors caused by noise in the telephone lines, ensuring the receipt of error-free files. The most common file transfer protocols are Kermit, Xmodem, Ymodem, and Zmodem.

III.   ARE YOU READY TO START SURFING THE NET?

The easiest and most popular way to navigate the Internet these days is the World Wide Web. The safety field has made its own inroads on the Web. A search on a specific topic, such as hearing protection, can yield numerous Internet sites. This would include companies that provide products for consumers, professional trade associations, journals and government regulations - all with a few clicks of the nouse.

A keyword search for "Safety" turned up 524 matches. Among the links one could follow were: "Business and Economy: Companies: Industrial Supplies: Safety Products," "Science: Agriculture: Food Safety," and "Recreation: Outdoors: Canoeing, Kayaking, and Rafting: Safety."

Internet users can choose from many search engines to conduct keyword searches. Search engines are cross-referenced indexes of millions of online documents. The most common Web-browsing search tools are *Yahoo!*, *Lycos* and *Alta Vista*. All users can access multiple search engines through their Internet providers.

To use search engines effectively, you must narrow your parameters. Broad single-word searches such as "Ergonomics" or "Chemicals" will turn up hundreds of entries. That's fine if you have the time to wander among the linked documents. It's wise, however, to look only at the first 10 to 20 documents offered, which are usually the most accurate "hits."

Or, you can make your search topics more specific. For example, if you pair-up keywords such as "Ergonomics" and "Vibration," or "Chemicals" and "Dermatitis," you will receive a more concise menu of documents.

The Web is growing by leaps and bounds every day. The safety field is certainly not to be left behind. Many government agencies can be accessed via the Web, numerous private companies have their own home pages, and many others are in the development stage.

1. The World Wide Web

The Web offers several significant features. First, it lets computers transmit not only text, but graphics and sounds, over the Internet. It also allows Internet Addresses, or links to other computers, to be embedded within a web page. (When navigating the World Wide Web, the files that you see on your computer screen when you connect to another computer are called *web pages*.)

My type of Internet connection lets me use web browser software, such as *Netscape* or *Microsoft*, that offers a graphical interface and supports the use of a mouse.

So by simply clicking on one of these links you can bring up another file stored on the current computer you will be connected to or open up a connection to another computer across the country.

2.   Sample Experience

To get an idea of how much information is available and how easy it is to locate, consider the path I took while just playing around. Let's start with OSHA's web page in Washington, D.C. OSHA's web page lets you view and download a copy of the draft ergonomics proposal, and check out other government Internet sites, or other safety and health links. A few more mouse clicks, and here's what I gained access to: EPA Chemical Substance Fact sheets from the University of Virginia, and Internet web sites for NIOSH, the National Library of Medicine, and Canadian Centre for Occupational Health and Safety. I also found links that would lead me to MSDS's online at the University of Utah, OSHA Federal Register notices, OSHA regulations, preamble to regulations, the field inspection reference manual (FIRM), corporate wide settlement agreements, and OSHA standards interpretations. Wow! That's a lot of information at your fingertips.

IV.   E-MAIL

You don't have to log onto the World Wide Web to gain access to helpful information. The Internet offers thousands of automated mailing lists, called list-servers, that are available for your subscription.

1.   Safety Mailing List

On the Safety mailing list, for instance, you can post a question one day and your message will arrive in the e-mail boxes of about 1,800 safety professionals the next day. More than likely, you'll receive several responses to your request.

There are a few things to keep in mind about this Safety mailing list. Because of the history of the Internet, 55 percent of the 1,800 subscribers are in educational institutions, and seventeen percent of subscribers are at companies; however, that number is rising. The mailing list only allows 50 messages per day to go out. The trick to avoiding e-mail overload is to simply delete messages based on their headings or titles, and open only those whose titles spark your interest.

When responding to someone else's posting, only send your response to the list when you think a group of people will be interested in your answer. Otherwise, send your response directly to the person's e-mail address.

The mailing lists that you can access via e-mail, and USENET newsgroups that you can access via newsreader software are as valuable as the search engines. (Most current Web-browser software provides access to both.)

A mailing list is an e-mail discussion group for people on a subscriber list run by Listserv software. Most mailing lists are free; you simply register electronically for a subscription. Any messages sent to the list are broadcast to all subscribers. To keep the flow of communication manageable, many subscribers use a digest option that collects the messages into large files and transmits them in bulk.

The best-known occupational safety and health mailing list is *SAFETY*. The address is: listserv@uvmvm.uvm.edu. Currently there are 2,300 subscribers on this list, with up to 50 messages posted daily. Another popular mailing list is the Duke University Occupational and Environmental Medicine site. The address is: occ-env-med-1@list.mc.duke.edu.

A newsgroup is more like an electronic bulletin board. Users can post a question or comment in a file in the hope that others will see it and respond.

2.   Easy Exploring

One of the benefits of the Web is the ease with which you can explore and dig up information. Consider the path I followed when I clicked on the phrase "Other safety and health links." This took me to the *Canadian Center for Occupational Health and Safety* whose computer is in Canada. There I found a link entitled *Computer-Related RSI primer*, which took me to a computer at the University of Nebraska.

This primer on computer-related ergonomic disorders included a general discussion of the hazards, as well as text and pictures on some stretches that could be used to address this problem. It also included a link with specific information on pointing devices, such as a computer mouse.

I followed this link to the University of California Berkeley where I came across some findings from a bioengineering Ph.D. candidate working at the ergonomics laboratory and conducting research on biochemical, physiological, and ergonomic factors surrounding pointing devices. I then jumped back to the computer-related RSI primer page because it included another link pointing to a typing injury resources list.

I followed this and found more than 50 links to documents containing information on topics such as keyboards for users with motion disabilities, tendinitis, dealing with insurance and lawyers, picking supports and splints, and how to correctly use arm rests. Using my browser software, which remembers all the sites I've visited, I then jumped back to the original OSHA page.

3. Conclusion

In evaluating this brief tour of the Internet, keep several points in mind:

- First, finding information can take some time. This brief journey took me about ten minutes, but there's no question that you can waste time exploring dead ends while browsing on the Web.

- On the other hand, by taking a few minutes to follow some links, you can stumble upon a gold mine of valuable data.

- Your web browser can save the address for a helpful site. So next time you log on, you need not start at OSHA's web page and follow ten links to reach that one resource you found helpful.

- You can locate information at web sites that offer databases and search engines that let you search by key word for a comprehensive listing of Internet sites addressing your topic. *Yahoo* is a search engine that is good to use when doing a keyword search.

- Keep in mind that because many web sites are new, web page designers are constantly updating and revising their information. Some web sites also become obsolete after not generating enough interest.

V. IMPORTANT INTERNET ADDRESSES TO START SURFING

To connect to a web site, you need to know the address of that computer on the Internet. We have provided some popular sites along with their addresses. Here's a small sampling of the information found on safety and health web pages and other interesting sites.

A little note to remember when surfing on the Net. When you find a site that you like and want to revisit, all Web-browser software has what is known as a "bookmark" that let's you save the web site. So the next time you log on, and you want to revisit that particular site, you will be able to pull it right up.

U. S. GOVERNMENT INTERNET SITES

1.   **OSHA**

2.   **Library of Congress**

3.   **White House Press Releases**

4.   **Executive Branch Resources**

5.   **NIOSH**

6.   **THOMAS**

7.   **U. S. House of Representatives**

8.   **U. S. Senate**

9.   **EPA**

10.   **The White House**

11.   **Bureau of Labor Statistics**

12.   **Federal Register**

13.   **FEDWORLD**

14.   **General Accounting Office**

15.   **Sen Edward Kennedy**

16.   **Supreme Court Decisions**

1. **OSHA - Occupational Safety and Health Administration**
   Visit OSHA at http://www.osha.gov/

   If you want to visit OSHA's web page, there are many area's to explore as you can see below. The "What's New" keeps you current with recent press releases and new standards in the works. It has a *Publications* section, *Frequently Asked Questions*, *Federal Register Notices* and much more.

2. **Library of Congress**
   Location: http://www.loc.gov/

   If the question is reference, here's the answer.

   The Library of Congress maintains millions of records of publications in the United States, as well as legislative and copyright information. The web site has pointers to the LC Marvel (the Library's database retrieval system). The page also has links to thousands of images from the American Memory project. Simply put, the Library of Congress is one of the best repositories on everything American.

3. **White House Press Releases**
   Location: http://www.ee.pdx.edu/cat/enchanter/html/wh.html

   Press releases and other information about White House characters.

4. **Executive Branch Resources via the Web**
   Location: http://lcweb.loc.gov/global/executive.html

   Never again will you have to hotfoot it around the Net looking for government information. The Library of Congress has compiled a web page that covers resources pertaining to the executive branch of the federal government and its various departments, as well as independent executive agencies.

5. **National Institute for Occupational Safety and Health**
   Visit NIOSH at http://www.cdc.gov/niosh/homepage.html

   Jump over to the NIOSH web page and download some fact sheets on indoor air quality, back belts, and carpal tunnel syndrome. You can also find a list of available publications, and information on how to get NIOSH training videos.

6.   **Thomas**
      Location: http://thomas.loc.gov/

Thomas provides the full text of bills as they make their way through the U.S. Senate and House of Representatives. Although the language of the bills can be arcane (you might need a lawyer to decode the subparagraphs), having this information freely available to the general public is an important feature of a democracy. You can choose to look at bills from the 104th Congress, and search for items based on the bill's popular title. Of special interest at Thomas is the link to the Congressional Record, where you can search the entire text of any newspaper or journal article that a member of Congress quoted when addressing an issue.

7.   **U. S. House of Representatives**
      Visit the House at http:/www.house.gov/

To get in touch with any member of the House of Representatives, visit this site and click on "Who's Who" and "How Do I Contact Them?" An alphabetical listing of members, committees, and subcommittees becomes available, complete with e-mail addresses for members who are technically avant garde. To view the House Calendar of sessions and recesses, simply click the Schedules section. This site also hosts general legal information on such topics as permanent laws in the United States and bills and legislation under consideration in Congress.

8.   **U. S. Senate**
          Gopher: //gopher.senate.gov:/

Tune in to the U. S. Senate page and dig up information on your senators and what duties they are performing as your representatives. Featured on-line are e-mail addresses for your senators so that you can contact them directly. Also available is a full file library including press releases.

9.   **Environmental Protection Agency**
          Visit EPA at http://www.epa.gov/

Visit EPA's web page, and you find information on press releases, free software, EPA initiatives, rules, regulations, and strategy documents, to name a few. A few mouse clicks will let you download a summary or the entire report for the recently announced Toxic Release Inventory data for 1993. Another link leads you to a pollution prevention directory that includes a list of resources such as clearinghouses, databases, periodicals and directories, hot lines, and an index of centers and associations. Or follow the links to EPA's Office of Solid Waste and Emergency Response and you'll find a copy of the agency's *RCRA Hazardous Waste Minimization Plan.*

10.   **The White House**
          Visit the White House at http://www.whitehouse.gov/

The White House's web page entitled welcome to the White House: an interactive citizen's handbook. Since its debut in October of 1994, an average of 10,000 people per hour have checked out President Clinton and Socks the Cat. There are many features to check out here. It has a history of the current First Family, White House tours, a welcome message from ... you guessed it! ... the President, publications, and a What's New section as well. You can even download an audio file of Socks the Cat meowing and send e-mail to the President. But don't forget to check out the audio message from Socks the Cat. Meow!

11.   Bureau of Labor Statistics
Visit BLS at http://stats.bls.gov/

The Bureau of Labor Statistics has a range of injury and illness information you can download. One link provides you with a list of contacts and phone numbers. Another fills your computer screen with the report of a 1992 worker injuries and illnesses by selected characteristics.

12.   Federal Registrar
Location: http://www.osha-slc.gov/toc_fed_reg.html

Documents from the Federal Register and information on how to gain access to the daily U. S. Federal Register via the Internet. Documents also include executive orders, and finding information on new OSHA regulations. You can follow the standard setting process in the Federal Register. A standard begins with publication in the Federal Register for a request for information (RFI), or an advance notice of proposed rulemaking (AN PRM), or a notice of proposed rulemaking (NPRM). OSHA seeks information to determine the extent of a particular hazard(s), current and potential protective measures, and the associated costs and benefits of various regulations.

13.   FedWorld
Location: http://www.fedworld.gov/

An enormous resource for scientific, technical, and other information provided by the federal government. FedWorld is taxpayer-supported through the National Technical Information Service (NTIS). It is an easy-to-use system that offers information on a wide variety of subjects.

14.   GAO (General Accounting Office) Reports
Gopher http://wiretap.spies.com

Reports from the GAO on budget issues, investment, OSHA, EPA, government management, public services, health care, energy issues, and virtually every other area of government is found here.

15. **Senator Edward Kennedy**
    Visit Teddy at http://www.ai.mit.edu/projects/iiip/Kennedy/homepage.html

    You don't have to be in Washington to march right into Ted Kennedy's office and see what's going on with the senator. Kennedy's electronic home away from home is now on the Web. Get the latest controversy that relates to Kennedy's office.

16. **Supreme Court Decisions**
    Visit the Court at http://www.law.cornell.edu/supct/

    Simply and clearly arranged, this site allows you to view Supreme Court rulings since 1990, by topic or from a searchable list. It's kept current when the Court is in session, and all opinions (both concurring and dissenting) are available. If you don't have a law degree, don't be intimidated. The site is very user- and modem-friendly, and includes explanations of legal terms and information about the way the Court works.

STATE LABOR DEPARTMENT WEB SITES

1. **Alaska Labor Standards and Safety Division**

2. **California Department of Industrial Relations**

3. **Hawaii Department of Labor and Industrial Relations**

4. **Maryland Division of Labor and Industry**

5. **Minnesota Department of Labor and Industry**

6. **New York Department of Labor**

7. **North Carolina Department of Labor**

8. **South Carolina Department of Labor**

9. **Tennessee Department of Labor**

10. **Utah Industrial Commission**

11. **VOSHA The Vermont Occupational Safety and Health Administration**

12. **Washington Department of Labor and Industries**

13. **Wyoming Department of Employment**

1.   **Alaska Labor Standards and Safety Division**
      Visit the home page at http://www.state.ak.us./local/akpageslabor/iss/iss.htm

      Read about all the state OSHA standards and the certification programs for
      asbestos abatement, hazardous painting, and explosives handling.

2.   **California Department of Industrial Relations**
      Visit the home page at: http://dir.ca.gov

      The Cal OSHA web site contains all the OSHA standards, news releases,
      apprenticeship courses to assist workers in job training, a mediation and
      conciliation department in addressing labor/management issues in the workplace
      and statistics on the labor force.

3.   **Hawaii Department of Labor and Industrial Relations**
      Visit the home page at http://hinc.hinc,hawaii.gov/hinc/

      Hawaii's web site offers all the OSHA standards and the latest news releases
      concerning job safety and health issues.

4.   **Maryland Division of Labor and Industry**
      Visit the home page at http://dllr.state.mo.us/dllr/thepage

      Maryland OSHA provides recent news releases, has a listing of all the MOSH
      standards, and offers information on unemployment insurance, job service, and job
      service Partnership Act. It has a division of Occupational and Professional
      Licensing and nuch more!

5.   **Minnesota Department of Labor and Industry**
      Visit its home page at http://dlli.state.mn.us

      Minnesota OSHA's web site offers a wealth of information. It has 10 different
      sections on the following:
      - *CompAct Newsletter*
      - Consultation Services

- Family Medical Leave
- Labor Laws
- Labor-Management
- Minimum Wage
- OSHA
- Safety Programs
- Vocational Rehab. & QRCs
- Work Comp FAQs

6. **New York Department of Labor**
   Visit its home page at http://labor.state.ny.us

New York OSHA's web site, like the Big Apple, is very interesting with lots of information. It has a section on what an employer needs to know in the state, publications on the workforce, and much more.

7. **North Carolina Department of Labor**
   Visit its home page at http://dol.state.nc.us/dol

North Carolina OSHA has a labor library service, a directory of labor officials, a monthly newsletter covering new developments on safety and health issues, a listing of publications, and you can review the latest Department of Labor News releases.

8. **South Carolina Department of Labor, Licensing & Regulation**
   Visit its home page at http://llr.sc.edu

South Carolina OSHA has a "What's New" section, an OSHA section, and an OSHA voluntary programs section.

9. **Tennessee Department of Labor**
   Visit its home page at http://inagural.state.tn.us/hp/sundquist/labor

Tennessee OSHA offers a News Release on the latest safety and health issues and all OSHA standards covered in their state.

10. **Utah Industrial Commision**
     Visit its home page at http://ind-com.state.ut.us

     Utah offers a newsletter, OSHA information and a section on OSHA books.

11. **Vermont Occupational Safety and Health Administration**
     Visit its home page at http://quest-net-com/vosha

     VOSHA offers answers to frequently asked questions and links to various safety and health sites.

12. **Washington Department of Labor and Industries**
     Visit its home page at http://wa.gov/lni

     Washington OSHA offers a "What's New" section, a section on press releases, a section about labor and industries, workers compensation insurance, industrial safety and health, specialty compliance (electrical training and contractor registration, OSHA standards, and a publications catalog.

13. **Wyoming Department of Employment**
     Visit its home page at http://www.state.wy.us/state/government/state-agencies/employment

     Wyoming OSHA offers: administration, labor standards, employment resources and a worker's safety and compensation section.

SAFETY & HEALTH WEB SITES

1. Agency for Toxic Substances and Disease Registry
2. American Industrial Hygiene Association (A.I.H.A.)
3. American Society of Safety Engineers (ASSE)
4. American National Standards Institute (ANSI)
5. Civil Engineering
6. Congressional E-Mail Directory
7. Dennison University, Campus Security & Safety
8. Duke University Occupational & Environmental Medicine
9. Emergency Preparedness Information Exchange (EPIX)
10. ERGO Web
11. Environmental Resource Center
12. Environmental Scorecard
13. MSDS On-Line
14. Michigan State University Radiation Chemical & Biological Safety
15. National Council on Compensation
16. National Safety Council
17. National Standards Systems Network (NSSN)
18. Rocky Mountain Center for Occupational & Environmental Health
19. Trench Safety
20. University of London Ergonomics & Human Computer Interaction
21. University of Virginia EPA Chemical Substance Factsheets
22. University of Virginia Display Ergonomics Page
23. Utah Safety Council
24. U. S. Dept. of Health and Human Services (HHS)
25. World Health Organization (WHO)

1.   **Agency for Toxic Substances and Disease Registry**
     Visit the agency at http://atsdrl.cdc.gov:8080/atsdrhome.html

     This agency has a listing of all the toxic chemicals. You can search the database to see if your company has any of these chemicals at your worksites.

2.   **American Industrial Hygiene Association**
     Visit its home page at http://www.aiha.org

     The American Industrial Hygiene Association (AIHA) posts information affecting the industrial hygiene profession.

3.   **American Society of Safety Engineers (ASSE)**
     Visit its home page at http://www.asse.org/

     The ASSE's home page contains membership information, educational offerings, technical publications, and government affairs information.

4.   **American National Standards Institute**
     Visit ANSI at: http://www.ansi.org/home.html

     ANSI is a private non-profit organization which means it does not write standards. It makes use of the combined technical talent and expertise of its organizational members, technical, professional and trade associations, as well as the companies and industrial firms that make up the organization. ANSI has been said to be the major clearinghouse and coordinating agency for voluntary standardization in the United States. ANSI has documents on any subject of safety. It covers everything. If you have a question about any standard - this is the site to use.

5.   **Civil Engineering**
     Location at http://akebono.stanford.edu/yahoo/science/engineering/civil-engineering/

     This web page has links to civil engineering resources including construction information, institutes, and water resources.

6. **Congressional E-Mail Directory**
   Location: http://astl.spa.umn.edu/juan/congress.html

This site, maintained by the University of Minnesota, makes it easy for you to participate in the legislative process by providing the e-mail addresses of all members of the Senate and House of Representatives. Just click the name of your state, territory, or district, and you'll find your elected officials' e-mail address, telephone number, and fax number. If you click the name of a state, you'll see its CapWeb page, where you'll find information about the committees on which the legislators from that state sit.

7. **Denison University, Campus Security & Safety**
   Location: http://www.denison.edu/sec-safe/

Denison has an employee safety training, laboratory safety, and a section containing government mandated safety plans.

8. **Duke University Occupational & Environmental Medicine**
   Location: http://occ-env-med.mc.duke.edu/oem

Duke's web page is very comprehensive. It is six pages long. It has many recent OSHA policies, like OSHA's guidelines for the prevention of workplace violence and NIOSH's proposal for a New Noise Standard. It also carries many of EPA's recent policies as well as occupational and environmental health topics.

9. **Emergency Preparedness Information EXchange (EPIX)**
   Gopher address: hoshi.cic.sfu.ca

The Emergency Preparedness Information EXchange is dedicated to the promotion of networking in support of disaster mitigation research and practice. It offers information on emergency and disaster management organizations, topics, conferences, and access to other emergency management resources.

10. **Ergo Web**
    Location: http://tucker.mech.utah.edu

    Ergo Web is the place for ergonomics on the World Wide Web. It offers volumes of useful ergonomic information for free to a large worldwide audience. It also offers subscription access to a sophisticated set of ergonomic job evaluation, analysis, design and redesign software. It is a one stop spot for ergonomics information, products, case studies, instructional materials, standards and guidelines, communication opportunities, ergonomics-related news, and more - all presented in an ergonomic format.

11. **Environmental Resource Center**
    Visit ERC at http://ftp.clearlake.ibm.com/erc/homepage.html

    The Environmental Resource Center (ERC) provides timely and cost effective access to environmental data and information.

12. **Environmental Scorecard**
    Visit Environmental Scorecard at:
    http://www.econet.apc.org/lcv/scorecard.html

    See how EcoNet rates your representatives' voting records on environmental issues. Use your mouse to select the state for which you would like to view environmental voting records. This page also has links to bill descriptions, information about the state of the enviremental movement, the League of Conservation voters, and EcoNet.

13. **MSDS On-line from University of Utah**
    gopher://atlas.chem.utah.edu:70/11/MSDS

    A complete listing of every MSDS sheet is here. It has a alphabetical index from A-Z on each MSDS sheet.

14. **MSU Radiation, Chemical & Biological Safety**
   Visit their home page at http://www.orcbs.msu.edu

   Michigan State's website offers a broad range of safety including the following: radiation safety, chemical safety, biological safety, hazardous waste and also offers training courses.

15. **National Council on Compensation**
   Visit NCCI's home page at http://focaraton.com/ncci/

   The NCCI provides free accident reduction and hazard control information.

16. **The National Safety Council**
   Visit their home page at http://www.ncs.org

   The NSC web page covers all the various categories of safety: safety and health, environmental, auto, home, traffic. It also has a facts and resources section, an international section, a publications section and a chapter news page.

17. **The National Standards Systems Network**
   Visit NSSN's home page at http://www.nssn.org/

   This site will be available in early 1997. The National Standards Systems Network (NSSN) will provide a wide range of standards information from major U.S. private sector standards developing organizations. It will have a catalog database to more than 60,000 standards. Users will be able to browse documents on-line, find out about standards that are in the process of being developed, and even comment on standards in the works.

18. **Rocky Mountain Center for Occupational and Environmental Health**
   Visit their home page at http://rocky.utah.edu

   The University of Utah is one of 14 NIOSH Educational Resource Centers. It covers industrial hygiene, occupational health nursing, air quality issues, occupational/environmental medicine, and ergonomics/safety issues.

The University of Utah is one of 14 NIOSH Educational Resource Centers. It covers industrial hygiene, occupational health nursing, air quality issues, occupational/environmental medicine, and ergonomics/safety issues.

19. **Trench Safety, A Tutorial for Constructors**
    Location at: http://www.auburn.edu/academic/
    architecture/bsc/research/trench/index.html

    This is Auburn Unversity's web page solely dedicated to trenching and shoring (cave in's) in the construction industry.

20. **University of London Ergonomics and Human Computer Interaction**
    Location at: http://www.ergohci.ucl.ac.uk/

    The Ergonomics and HCI Unit is UCL's centre for the development, study and practice of ergonomics and human-computer interaction. Ergonomics is concerned with the human aspects of technology at work. It seeks to ensure that technology meets the needs of people that use it. Human-Computer Interaction is a field of ergonomics that is particularly concerned with computer technology. The unit's teaching group offers a one year MSC taught course in Ergonomics. Studies in ergonomics may also be pursued at doctoral level.

21. **University of Virginia EPA Chemical Substance Factsheets**
    gopher://ecosys.drdr.virginia.edu/11/library/gen/toxics

    Information from the EPA on hundreds of chemicals, elements, and compounds. Factsheets include data on toxicity, identification, reason for citation, how to determine if you've been exposed, OSHA safety limits, ways to reduce exposure, and more relevant information. The factsheets are alphabetical from A-Z.

22. **Utah Safety Council**
    Location: http://www.ps.ex.state.ut.us/sc/usc/htm

    The Utah Safety Council's web page covers a safety video 'library," defensive driving course, recreational safety, and much more.

23. **UVA's Video Display Ergonomics page**
     Location at: http://www.virginia.edu/~enhealth/ergonomics/toc.html

The University of Virginia web page covers Ergonomics. It covers Office Ergonomics on video display terminals,computer stretch breaks and back injury prevention on how to avoid a painful back.

24. **U.S. Department of Health and Human Services**
     Visit it's home page at http://www.os.dhhs.gov

The HHS web page covers public affairs, what's new, consumer information, and a whole lot more.

25. **World Health Organization**
     Visit its home page at http://www.who.ch.

The WHO web page is for anyone who wants to know more about world health issues. It has a newsletter, publications, and a WHO Library.

APPENDIX A

SELF-INSPECTION CHECKLISTS

A Do-It-Yourself OSHA Inspection Checklist

One of the major challenges in complying with the Occupational Safety and Health Act is determining which of the thousands of regulations apply to you, and where to start in bringing your shop up to standards.

There is no easy answer, but the attached safety check sheet is an excellent starting point. It calls your attention to many of the major safety points that an OSHA Inspector will be looking for when he inspects your plant. By using this check sheet to conduct your own plant inspection before the OSHA man arrives, you will be alerted to many of the potential violations that could result in stiff fines. Correcting any unsafe conditions will be your next objective, but you will at least know what has to be done.

Self-inspection Checklists

Employer Posting

☐ Is the OSHA poster, *Safety and Health Protection on the Job*, displayed in a prominent location where all employees are likely to see it?

Are other posters or notices properly or displayed, such as:

☐ OSHA 200 Summary in February?

☐ Are emergency telephone numbers posted where they can be readily used in case of emergency?

☐ Where employees may be exposed to any toxic substances or harmful physical agents, has appropriate information concerning employee access to medical and exposure records and Material Safety Data Sheets (MSDSs) been posted or otherwise made readily available to affected employees?

☐ Are signs regarding exits from buildings, room capacity, floor loading, exposure to x-ray, micro-wave, or other harmful radiation or substances posted where required?

Recordkeeping

☐ Are all occupational injuries & illnesses, except minor injuries requiring only first aid, being recorded as required on the *OSHA Form 200*?

☐ Are copies of *OSHA Form 200* and First Report of Injury, Form 101, kept for five years?

☐ Are employee medical records and records of employee exposures to hazardous substances or harmful physical agents current?

☐ Have arrangements been made to maintain required records for the legal period of time for each specific type of record? (Some records must be maintained for at least 40 years.)

☐ Are operating permits and records current for such items as elevators, pressure vessels, and liquefied petroleum gas tanks?

☐ Are employee safety and health training records maintained?

☐ Is documentation of safety inspections and corrections maintained?

Injury & Illness Prevention Program

☐ Do you have top management commitment?

☐ Do you have a system in place for hazard identi-fication and control?

☐ Have you established labor and management accountability?

☐ Do you investigate all incidents and accidents?

☐ Do you encourage employee involvement in health and safety matters?

☐ Do you provide occupational safety and health training for your workers and supervisors?

☐ Do you perform periodic evaluations of the program?

Medical Services & First Aid

☐ Has an emergency medical plan been developed?

☐ Are emergency phone numbers posted?

☐ Are first aid kits easily accessible to each work area, with necessary supplies available, periodi-cally inspected and replenished as needed?

☐ Are means provided for quick drenching or flushing of the eyes and body in areas where caustic or corrosive liquids or materials are handled?

Safety Committees

☐ Do you have an active safety committee with equal numbers of management and employees?

☐ Are records kept documenting safety and health training for each employee by name or other identifier, training dates, type(s) of training, and training provider?

☐ Does the committee meet at least monthly?

☐ Is a written record of safety committee meetings distributed to affected employees, and maintained for division review?

☐ Does the safety committee conduct quarterly hazard identification surveys?

☐ Does the committee review results of periodic, scheduled work site inspections?

☐ Does the committee review accident and near-miss investigations and, where necessary, submit recommendations for prevention of future incidents?

☐ Does the committee involve all workers in the safety and health program?

☐ Are safety committee minutes kept three years and are each month's minutes posted?

☐ Has your safety committee developed an accident investigation procedure?

☐ Has the committee reviewed your safety and health program and made recommendations for possible improvements?

☐ Have committee members been trained and instructed in safety committee purpose and operation, methods of conducting meetings, OSHA rules which apply to the workplace, hazard identification, and accident investigation principles?

Fire Protection

☐ Do you have a written fire-prevention plan?

☐ Does your plan describe the type of fire protection equipment and/or systems?

☐ Have you established practices and procedures to control potential fire hazards and ignition sources?

☐ Are employees aware of the fire hazards of the materials and processes to which they are exposed?

☐ Is your local fire department well acquainted with your facilities, location, and specific hazards?

☐ If you have a fire alarm system, is it tested at least annually?

☐ Are sprinkler heads protected by metal guards when exposed to physical damage?

☐ Is proper clearance maintained below sprinkler heads?

☐ Are portable fire extinguishers provided in adequate numbers and types?

☐ Are fire extinguishers mounted in readily assessable locations?

☐ Are fire extinguishers recharged regularly and then noted on the inspection tag?

☐ Are employees trained in the use of extinguishers and fire protection procedures?

Personal Protective Equipment & Clothing

☐ Are protective goggles or face shields provided and worn where there is any danger of flying particles or corrosive materials??

☐ Are approved safety glasses required to be worn at all times in areas where there is risk of eye injury such as punctures, abrasions, contusions, or burns?

☐ Are protective gloves, aprons, shields or other protection provided against cuts, corrosive liquids, and chemicals?

☐ Are hard hats provided and worn where danger of flying or falling objects exists?

☐ Are hard hats inspected periodically for damage to the shell and suspension system?

☐ Is appropriate foot protection required where there is risk of foot injury from hot, corrosive, poisonous substances, falling objects, crushing, or penetrating actions?

☐ Are approved respirators provided for regular or emergency use where needed?

☐ Is all protective equipment maintained in a sanitary condition and ready for use?

☐ Do you have eyewash facilities and a quick-drench shower within a work area where employees are exposed to caustic or corrosive materials?

☐ When lunches are eaten on the premises, are they eaten in areas where there is no exposure to toxic materials or other health hazards?

☐ Is protection against the effects of occupational noise exposure provided when sound levels exceed those of the OSHA noise and hearing conservation standard?

General Work Environment

☐ Are all work sites clean and orderly?

☐ Are work surfaces kept dry or appropriate means taken to assure the surfaces are slip-resistant?

☐ Are all spilled materials or liquids cleaned up immediately?

☐ Is combustible scrap, debris, and waste stored safely and removed from the work site promptly?

☐ Are covered metal waste cans used for oily and paint-soaked waste?

☐ Are the minimum number of toilets and washing facilities provided?

☐ Are all toilets and washing facilities clean and sanitary? Are all work areas adequately lighted?

Walkways

☐ Are aisles and passageways kept clear and are they at least 22 inches wide?

☐ Are aisles and walkways appropriately marked? Are wet surfaces covered with non-slip materials?

☐ Are openings or holes in the floors or other treading surfaces repaired or otherwise made safe?

☐ Is there safe clearance for walking in aisles where vehicles are operating?

☐ Are materials or equipment stored so sharp objects can not obstruct the walkway?

☐ Are changes of direction or elevations readily identifiable?

☐ Are aisles or walkways that pass near moving or operating machinery, welding operations, or similar operations arranged so employees will not be subjected to potential hazards?

☐ Is adequate headroom (of at least 6.5 feet) provided for the entire length of any walkway?

☐ Are standard guardrails provided wherever aisle or walkway surfaces are elevated more than four feet above any adjacent floor or the ground?

☐ Are bridges provided over conveyors and similar hazards?

Floor & Wall Openings

☐ Are floor holes or openings guarded by a cover, guardrail, or equivalent on all sides (except at entrance to stairways or ladders)?

☐ Are toe boards installed around the edges of a permanent floor opening (where persons may pass below the opening)?

☐ Are skylight screens of such construction and mounting that they will withstand a load of at least 200 pounds?

☐ Is the glass in windows, doors, and glass walls (which may be subject to human impact) of sufficient thickness and type for all conditions of use?

☐ Are grates or similar covers over floor openings, such as floor drains, of such design that foot traffic or rolling equipment will not be caught by the grate spacing?

☐ Are unused portions of service pits and pits not actually in use either covered or protected by guardrails or equivalent?

Stairs & Stairways

☐ Are standard stair rails and handrails present on all stairways having four or more risers?

☐ Are all stairways at least 22 inches wide?

☐ Do stairs have at least a 6.5 feet overhead clearance?

☐ Do stairs angle no more than 50 degrees and no less than 30 degrees?

☐ Are step risers on stairs uniform from top to bottom, with no riser spacing greater than 7.5 inches?

☐ Are steps on stairs and stairways designed or provided with a surface that renders them slip resistant?

☐ Are stairway handrails located between 30-34 inches above the leading edge of stair treads?

☐ Do stairway handrails-have at least 1.5 inches clearance between handrails and the wall or surface they are mounted on?

☐ Are stairway handrails capable of withstanding a load of 200 pounds applied in any direction?

☐ Where stairs or stairways exit directly into any area where vehicles may be operated, are adequate barriers and warnings provided to prevent employees from stepping into the path of traffic?

Elevated Surfaces

☐ Are signs posted, when appropriate, showing elevated floor load capacity?

☐ Are elevated surfaces (more than four feet above the floor or ground) provided with standard guardrails?

☐ Are all elevated surfaces (beneath which people or machinery could be exposed to falling objects) provided with standard toe boards?

☐ Is a permanent means of access/egress provided to elevated work surfaces?

☐ Is material on elevated surfaces piled, stacked, or racked in a manner to prevent it from tipping, falling, collapsing, rolling, or spreading?

☐ Are dock boards or bridge plates used when transferring materials between docks and trucks or railcars??

☐ When in use, are dock boards or bridge plates secured in place?

Exit or Egress

☐ Are all exits marked with an exit sign and illuminated by a reliable light source?

☐ Are the directions to exits, if not immediately apparent, marked with visible signs?

☐ Are doors, passageways, or stairways, that are neither exits nor access to exits and which could be mistaken for exits, appropriately marked "NOT AN EXIT," or "TO BASEMENT," "STOREROOM" and the like?

☐ Are exit signs provided with the word "EXIT" in lettering at least five inches high and the stroke of the lettering at least 1/2 inch wide?

☐ Are exit doors side-hinged? Are all exits kept free of obstructions and unlocked?

☐ Are at least two means of egress provided from elevated platforms, pits or rooms where the absence of a second exit would increase the risk of injury from hot, poisonous, corrosive, suffo-cating, flammable, or explosive substances?

☐ Are there sufficient exits to permit prompt escape in case of emergency?

☐ Are the number of exits from each floor of a building and the number of exits from the building itself appropriate for the building occupancy load?

☐ When workers must exit through glass doors, storm doors and such, are the doors fully tempered and meeting safety requirements for human impact?

Exit Doors

☐ Are doors which are required to serve as exits designed and constructed so that the way of exit travel is obvious and direct?

☐ Are windows (which could be mistaken for exit doors) made inaccessible by barriers or railing?

☐ Are exit doors able to open from the direction of exit travel without the use of a key or any special knowledge or effort?

☐ Is a revolving, sliding, or overhead door prohibited from serving as a required exit door?

☐ When panic hardware is installed on a required exit door, will it allow the door to open by applying a force of 15 pounds or less in the direction of the exit traffic?

☐ Are doors on cold-storage rooms provided with an inside release mechanism which will release the latch and open the door even if it is padlocked or otherwise locked on the outside?

☐ Where exit doors open directly onto any street, alley, or other area where vehicles may be operated, are adequate barriers and warnings provided to prevent employees from stepping directly into the path of traffic?

☐ Are doors that swing in both directions and are located between rooms where there is frequent traffic, provided with viewing panels in each door?

Portable Ladders

☐ Are all ladders maintained in good condition, joints between steps and side rails tight, all hardware and fittings securely attached, and moveable parts operating freely without binding or undue play?

☐ Are nonslip safety feet provided on each ladder including metal or rung ladders?

☐ Are ladder rungs and steps free of grease and oil?

☐ Is it prohibited to place a ladder in front of doors opening toward the ladder except when the door is blocked open, locked, or guarded?

☐ Is it prohibited to place ladders on boxes, barrels, or other unstable bases to obtain additional height?

☐ Are employees instructed to face the ladder when ascending / descending?

☐ Are employees prohibited from using ladders that are broken, missing steps, rungs or cleats, broken side rails, or other faulty equipment?

☐ Are employees instructed not to use the top step of ordinary stepladders as a step?

☐ When portable rung ladders are used to gain access to elevated platforms, roofs, and the like, does the ladder always extend at least three feet above the elevated surface?

☐ Is it required that when portable rung or cleat-type ladders are used, the base is so placed that slipping will not occur, or it is lashed or otherwise held in place?

☐ Are portable metal ladders legibly marked with signs reading "CAUTION-Do Not Use Around Electrical Equipment" or equivalent wording?

☐ Are the rungs of ladders uniformly spaced at 12 inches, center to center?

Hand Tools & Equipment

☐ Are all tools and equipment (both company and employee-owned) in good working condition?

☐ Are hand tools such as chisels or punches (which develop mushroomed heads during use) reconditioned or replaced as necessary?

☐ Are broken or fractured handles on hammers, axes, or similar equipment replaced promptly?

☐ Are appropriate handles used on files and similar tools?

☐ Are appropriate safety glasses, face shields, and similar equipment used while using hand tools or equipment which might produce flying materials or be subject to breakage?

☐ Are jacks checked periodically to assure that they are in good operating condition?

☐ Are tool handles wedged tightly in the head of all tools?

☐ Are tool-cutting edges kept sharp so the tool will move smoothly without binding or skipping?

☐ Is eye and face protection used when driving hardened or tempered tools, bits, or nails?

Portable (Power-Operated) Tools & Equipment

☐ Are grinders, saws, and similar equipment provided with appropriate safety guards?

☐ Are power tools used with the shield or guard recommended by the manufacturer?

☐ Are portable circular saws equipped with guards above and below the base shoe?

☐ Are circular saw guards checked to assure guarding of the lower blade portion?

☐ Are rotating or moving parts of equipment guarded to prevent physical contact?

☐ Are all cord-connected, electrically-operated tools and equipment effectively grounded or of the approved double-insulated type?

☐ Are effective guards in place over belts, pulleys, chains, and sprockets on equipment such as concrete mixers, air compressors, and the like?

☐ Are portable fans provided with full guards having openings of 1/2 inch or less?

☐ Is hoisting equipment available and used for lifting heavy objects, and are hoist ratings and characteristics appropriate for the task?

☐ Are ground-fault circuit interrupters (provided on all temporary electrical 15 and 20 ampere circuits) used during periods of construction?

☐ Are pneumatic and hydraulic hoses on power-operated tools checked regularly for deterioration or damage?

Abrasive Wheel Equipment Grinders

☐ Is the work rest used and kept adjusted to within 1/8 inch of the wheel?

☐ Is the adjustable tongue on the top side of the grinder used and kept adjusted to within 1/4 inch of the wheel?

☐ Do side guards cover the spindle, nut, flange, and 75 percent of the wheel diameter?

☐ Are bench and pedestal grinders permanently mounted?

☐ Are goggles or face shields always worn when grinding?

☐ Is the maximum RPM rating of each abrasive wheel compatible with the RPM rating of the grinder motor?

☐ Are fixed or permanently mounted grinders connected to their electrical supply system with metallic conduit or by another permanent wiring method?

☐ Does each grinder have an individual on/off switch?

☐ Is each electrically-operated grinder effectively grounded?

☐ Before mounting new abrasive wheels, are they visually inspected and ring tested?

☐ Are dust collectors and powered exhausts provided on grinders used in operations that produce large amounts of dust?

☐ To prevent coolant from splashing workers, are splash guards mounted on grinders that use coolant?

☐ Is cleanliness maintained around grinders?

Machine Guarding

☐ Is there an employee training program for safe methods of machine operation?

☐ Is there adequate supervision to ensure that employees are following safe machine operating procedures?

☐ Is there a regular program of safety inspection for machinery and equipment?

☐ Is all machinery and equipment clean and properly maintained?

☐ Is sufficient clearance provided around and between to allow for safe operations, set up servicing, material handling, and waste removal?

☐ Is equipment and machinery securely placed and anchored when necessary to prevent tipping or other movement that could result in personal injury?

☐ Is there a power shut-off switch within reach of the operator's position at each machine?

☐ Are the noncurrent-carrying metal parts of electrically-operated machines bonded and grounded?

☐ Are foot-operated switches guarded or arranged to prevent accidental actuation by personnel or falling objects?

☐ Are manually-operated valves and switches (controlling the operation of equipment and machines) clearly identified and readily accessible?

☐ Are all emergency-stop buttons colored red?

☐ Are all pulleys and belts (that are located within seven feet of the floor or working level) properly guarded?

☐ Are methods provided to protect the operator and other employees in the machine area from hazards created at the point of operation, ingoing nip points, rotating parts, flying chips, and sparks?

☐ Are all moving chains and gears properly guarded?

☐ Are machinery guards secured and arranged so they do not offer a hazard in their use?

☐ If special hand tools are used for placing and removing material, do they protect the operator's hands?

☐ Are revolving drums, barrels, and containers (required to be guarded by an enclosure that is interlocked with the drive mechanism so that revolution cannot occur) guarded?

☐ Do arbors and mandrels have firm and secure bearings, and are they free from play?

☐ Are provisions made to prevent machines from automatically starting when power is restored (following a power failure or shut-down)?

☐ Are machines constructed so as to be free from excessive vibration (when the largest size tool is mounted and run at full speed)?

☐ If machinery is cleaned with compressed air, is air pressure controlled and personal protective equipment or other safeguards used to protect operators and other workers from eye and body injury?

☐ Are fan blades protected with a guard having openings no larger than 1/2 inch when operating within seven feet of the floor?

☐ Are saws used for ripping equipped with anti-kickback devices and spreaders?

☐ Are radial arm saws guarded and so arranged that the cutting head will gently return to the back of the table when released?

Lockout/Tagout Procedures

☐ Is all machinery or equipment (capable of movement) required to be de-energized or disengaged and locked out during cleaning, servicing, adjusting, or setting-up operations?

☐ Is it prohibited to lock-out control circuits in lieu of locking-out main power disconnects?

☐ Are all equipment control valve handles provided with a means of lock out?

☐ Does the lockout/tagout procedure require that stored energy (i.e., mechanical, hydraulic, air) be released or blocked before equipment is locked out for repairs?

☐ Are appropriate employees provided with individually keyed personal safety locks?

☐ Are employees required to keep personal control of their key(s) while they have safety locks in use?

☐ Is it required that employees check the safety of the lockout by attempting to start up after making sure no one is exposed?

Where the power disconnecting means for equipment does not also disconnect the electrical control circuit:

☐ Are the appropriate electrical enclosures identified?

☐ Are means provided to assure the control circuit can also be disconnected and locked out?

Welding, Cutting & Brazing

☐ Are only authorized and trained personnel permitted to use welding, cutting, or brazing equipment?

☐ Are compressed gas cylinders regularly examined for signs of defect, deep rusting, or leakage?

☐ Are cylinders kept away from sources of heat? Is it prohibited to use cylinders as rollers or supports?

☐ Are empty cylinders appropriately marked, their valves closed, and valve-protection caps placed on them?

☐ Are signs reading: "DANGER - NO SMOKING, MATCHES OR OPEN LIGHTS," or the equivalent posted?

☐ Are cylinders, cylinder valves, couplings, regulators, hoses, and apparatus kept free of oily or greasy substances?

☐ Unless secured on special trucks, are regulators removed and valve-protection caps put in place before moving cylinders?

☐ Do cylinders without fixed hand wheels have keys, handles, or nonadjustable wrenches on stem valves when in service?

☐ Are liquefied gases stored and shipped with the valve end up, and with valve covers in place?

☐ Before a regulator is removed, is the valve closed and then gas released from the regulator?

☐ Is open circuit (no load) voltage of arc welding and cutting machines as low as possible, and not in excess of the recommended limit?

☐ Are electrodes removed from the holders when not in use?

☐ Is it required that electric power to the welder be shut off when no one is in attendance?

☐ Is suitable fire extinguishing equipment available for immediate use?

☐ Is welder forbidden to coil or loop welding electrode cable around his/her body?

☐ Are work and electrode lead cable frequently inspected for wear and damage, and replaced when needed?

☐ Do means for connecting cable lengths have adequate insulation?

☐ When the object to be welded cannot be moved and fire hazards cannot be removed, are shields used to confine heat, sparks, and slag?

☐ Are fire watchers assigned when welding or cutting is performed in locations where a serious fire might develop?

☐ When welding is done on all metal walls, are precautions taken to protect combustibles on the other side?

☐ Before hot work begins, are drums, barrels, tanks, and other containers so thoroughly cleaned and tested that no substances remain that could explode, ignite, or produce toxic vapors?

☐ Do eye protection helmets, hand shields, and goggles meet appropriate standards?

☐ Are employees exposed to the hazards created by welding, cutting, or brazing operations protected with personal protective equipment and clothing?

☐ Is a check made for adequate ventilation in and where welding or cutting is performed?

☐ When working in confined spaces, are environmental monitoring tests taken and means provided for quick removal of welders in case of an emergency?

Compressors & Compressed Air

☐ Are compressors equipped with pressure-relief valves and pressure gauges?

☐ Are compressor air intakes installed and equipped to ensure that only clean, uncontaminated air enters the compressor?

☐ Are air filters installed on the compressor intake?

☐ Are compressors operated and lubricated in accordance with the manufacturer's recommendations?

☐ Are safety devices on compressed air systems checked frequently?

☐ Before any repair work is done on the pressure systems of the compressor, is the pressure bled off and the system locked out?

☐ Are signs posted to warn of the automatic starting feature of the compressors?

☐ Is the belt drive system totally enclosed to provide protection in the front, back, top, and sides?

☐ Is it strictly prohibited to direct compressed air toward a person?

☐ Are employes prohibited from using compressed air (at over 29 PSI) for cleaning purposes?

☐ Are employees prohibited from cleaning off clothing with compressed air?

☐ When using compressed air for cleaning, do employees use personal protective equipment?

☐ Are safety chains or other suitable locking devices used at couplings of high pressure hose lines where a connection failure would create a hazard?

☐ Before compressed air is used to empty containers of liquid, is the safe working pressure of the container checked?

☐ When compressed air is used with abrasive blast cleaning equipment, is the operating valve a type that must be held open manually?

☐ When compressed air is used with abrasive blast cleaning equipment, is the operating valve a type that must be held open manually?

☐ When compressed air is used to inflate auto tires, is a clip-on chuck (and an inline regulator preset to 40 p.s.i.) required?

☐ Is it prohibited to use compressed air to clean up or move combustible dust, if such action could cause the dust to be suspended in the air and cause a fire or explosion?

☐ If plastic piping is used, is the plastic approved for air line service? (ABS is OK - PVC is not.)

Compressed Gas & Cylinders

☐ Are cylinders with water-weight capacity over 30 pounds of equipped (with means for connecting a valve protector or device, or with a collar or recess) to protect the valve?

☐ Are cylinders legibly marked to clearly identify the gas contained?

☐ Are compressed gas cylinders stored in areas which are protected from external heat sources (such as flames impingement, intense radiant heat, electric arcs or high temperature lines)?

☐ Are cylinders stored or transported in a manner to prevent them from creating a hazard by tipping, falling, or rolling?

☐ Are cylinders containing liquefied fuel gas stored or transported in a position so that the safety relief device is always in direct contact with the vapor space in the cylinder?

☐ Are valve protectors always placed on cylinders when the cylinders are not in use or connected for use?

☐ Are the valves closed off before a cylinder is moved, when the cylinder is empty, and at the completion of each job?

☐ Are low-pressure fuel-gas cylinders checked periodically for corrosion, general distortion, cracks or any other defect that might indicate a weakness or render them unfit for service?

☐ Does the periodic check of low-pressure fuel gas cylinders include a close inspection of the bottom of each cylinder?

Industrial Trucks - Forklifts

☐ Are only trained personnel allowed to operate industrial trucks?

☐ Is substantial overhead protective equipment provided on high-lift rider equipment?

☐ Are the required lift-truck operating rules posted and enforced and is the capacity rating posted in plain view of the operator?

☐ Is directional lighting provided one each industrial truck that operates in an area with less than two foot-candles per square foot of general lighting?

☐ Does each industrial truck have a warning horn, whistle, gong, or other device which can be clearly heard above the normal noise in the area where operated?

☐ Are the brakes on each industrial truck capable of bringing the vehicle to a complete and safe stop when fully loaded?

☐ Will the industrial truck's parking brake effectively prevent the vehicle from moving when unattended?

☐ Are industrial trucks operating in areas where flammable gases or vapors, combustible dust, or ignitable fibers may be present in the atmosphere, approved for such locations?

☐ Are motorized hand and hand/rider trucks so designed that the brakes are applied and power to the drive motor shuts off when the operator releases his/her grip on the device that controls the travel?

☐ Are industrial trucks with internal combustion engines (and operated in building or enclosed areas) carefully checked to ensure such operations do not cause harmful concentrations of dangerous gases or fumes?

Spray Finishing Operations

☐ Is adequate ventilation assured before spray operations are started?

☐ Is mechanical ventilation provided when spraying is performed in enclosed areas?

☐ When mechanical ventilation is provided during spraying operations, is it arranged so that it will not circulate contaminated air?

☐ Is the spray area free of hot surfaces?

☐ Is the spray area at least 20 feet from flames, sparks, operating electrical motors, and other ignition sources?

☐ Are the portable lamps used to illuminate spray areas suitable for use in a hazardous location?

☐ Is approved respiratory equipment provided and used during spraying operations?

☐ Do solvents used for cleaning have a flash point of 100° F or more?

☐ Are fire control sprinkler heads kept clean?

☐ Are "NO SMOKING" signs posted in the spray areas, paint rooms, paint booths, and paint storage areas?

☐ Is the spray area kept clean of combustible residue?

☐ Are spray booths constructed of metal, masonry, or other substantial noncombustible material?

☐ Are spray booth floors and baffles noncombustible and easily cleaned?

☐ Is infrared drying apparatus kept out of the spray area during spraying operations?

☐ Is the spray booth completely ventilated before the drying apparatus is used? Is the electric drying apparatus properly grounded? Do all drying spaces have adequate ventilation?

☐ Are lighting fixtures for spray booths located outside the booth, and the interior lighted through sealed clear panels?

☐ Are the electric motors for exhaust fans placed outside booths or ducts?

☐ Are belts and pulleys inside the booth fully enclosed?

☐ Do ducts have access doors to allow cleaning?

Confined Spaces

☐ Is there a written permit confined space program?

☐ Is the program available for inspection?

☐ Are confined spaces thoroughly emptied of any corrosive or hazardous substances, such as acids or caustics, before entry?

☐ Before entry, are all pipe lines to a confined space containing inert, toxic, flammable, or corrosive materials valved off and blanked or disconnected and separated?

☐ Are all impellers, agitators, or other moving equipment inside confined spaces locked out if they present a hazard?

☐ Is either natural or mechanical ventilation provided prior to confined space entry?

☐ Before entry, are appropriate atmospheric tests performed to check for oxygen deficiency, toxic substances, and explosive concentrations in the confined space?

☐ Is adequate lighting provided for the work being performed in the confined space?

☐ Is the atmosphere inside the confined space frequently tested or continuously monitored during the work process?

☐ Is there an attendant standing by outside the confined space, whose sole responsibility is to watch the work in progress, sound an alarm if necessary, and help render assistance?

☐ Is the attendant or other employees prohibited from entering the confined space without lifelines and respiratory equipment if there is an emergency?

☐ In addition to the attendant, is there at least one other trained rescuer in the vicinity?

☐ Are all rescuers appropriately trained and using approved, recently inspected equipment?

☐ Does all rescue equipment allow for lifting employees vertically through a top opening?

☐ Are rescue personnel first aid and CPR-trained and immediately available?

☐ Is there an effective communication system in place whenever respiratory equipment is used, and the employee in the confined space is out of sight of the attendant?

☐ Is approved respiratory equipment required if the atmosphere inside the confined space cannot be made acceptable?

☐ Is all portable electrical equipment used inside confined spaces either grounded and insulated or equipped with ground-fault protection?

☐ Before gas welding or burning is started in a confined space, are hoses checked for leaks, compressed gas bottles forbidden inside the confined space area, and the confined space area tested for an explosive atmosphere each time before a lighted torch is taken into the confined space?

☐ When using oxygen-consuming equipment (such as salamanders, torches, furnaces) in a confined space, is air provided to assure combustion of the atmosphere below 19.5 percent by volume?

☐ Whenever combustion-type equipment is used in a confined space, are provisions made to ensure that the exhaust gases are vented outside the enclosure?

☐ Is each confined space checked for decaying vegetation or animal matter which may produce methane?

☐ Is the confined space checked for possible industrial waste which could contain toxic properties?

☐ If the confined space is below the ground and near areas where motor vehicles are operating, is it possible for vehicle exhaust or carbon monoxide to enter the space?

Environmental Controls

☐ Are all work areas properly lighted?

☐ Are hazardous substances identified which may cause harm by inhalation, ingestion, skin absorption, or contact?

☐ Are employees aware of the hazards involved with the various chemicals they may be exposed to in their work environment, such as ammonia, chlorine, epoxies, and caustics?

☐ Is employee exposure to chemicals in the workplace kept within acceptable levels? Can a less harmful method or product be used?

☐ Is the work area's ventilation system appropriate for the work being performed?

☐ Are proper precautions being taken when handling asbestos and other fibrous materials?

☐ Is the possible presence of asbestos determined prior to the beginning of any repair, demolition, construction, or reconstruction work?

☐ Are asbestos covered surfaced kept in good repair to prevent release of fibers?

☐ Are wet methods used (when predictable) to prevent emission or airborne asbestos fibers, silica dust, and similar hazardous materials?

☐ Is vacuuming with appropriate equipment conducted, rather than blowing or sweeping dust?

☐ Are grinders, saws, and other machines that produce respirable dusts vented to an industrial collector or central exhaust system?

☐ Are all local exhaust ventilation systems designed and operated properly (at the airflow and volume necessary) for the application? Are the ducts free of obstructions? Have you checked to ensure that the belts are not slipping?

☐ Is personal protective equipment provided, used, and maintained whenever required?

☐ Are there written *Standard Operating Procedures* for the selection and use of respirators?

☐ Are restrooms and washrooms kept clean and sanitary?

☐ Is all water (provided for drinking, washing, and cooking) potable?

☐ Are all outlets for water (that is not suitable for drinking) clearly identified?

☐ Are employees instructed in the proper manner of lifting heavy objects?

☐ Where heat is a problem, have all fixed work areas been provided with proper means of cooling?

☐ Are employees working on streets and roadways, where they are exposed to the hazards of traffic, required to wear high-visibility clothing?

☐ Are exhaust stacks and air intakes located so that contaminated air will not be recirculated within a building or other enclosed area?

Flammable & Combustible Materials

☐ Are combustible scrap, debris, and waste materials stored in covered metal receptacles, and remove from the work site promptly?

☐ Are proper storage methods used to minimize the risk of fire and spontaneous combustion?

☐ Are approved containers and tanks used for the storage and handling of flammable and combustible liquids?

☐ Are all connections on drums and combustible liquid piping (vapor and liquid) tight?

☐ Are all flammable liquids kept in closed containers when not in use?

☐ Are bulk drums of flammable liquids grounded and bonded to containers during dispensing?

☐ Do storage rooms for flammable and combustible liquids have explosion-proof lights?

☐ Do storage rooms for flammables and combustible liquids have mechanical or gravity ventilation?

☐ Are safe practices followed when liquid petroleum gas is stored, handled, and used?

☐ Are liquefied petroleum storage tanks guarded to prevent damage from vehicles?

☐ Are all solvent wastes and flammable liquids kept in fire-resistant, covered containers until they are removed from the work site?

☐ Is vacuuming used whenever possible, rather than blowing or sweeping combustible dust?

☐ Are fire separators placed between containers of combustibles or flammables when stacked one upon another (to assure their support and stability)?

☐ Are fuel-gas cylinders and oxygen cylinders separated by distance, fire-resistant barriers or other means while in storage?

☐ Are fire extinguishers provided for the type of materials they will extinguish, and placed in areas where they are to be used?

CLASS A: *Ordinary combustible materials fires*
CLASS B: *Flammable liquid, gas, or grease fires*
CLASS C: *Energized-electrical equipment fires*

☐ If a Halon 1301 fire extinguisher is used, can employees evacuate within the specified time (for that extinguisher)?

☐ Are appropriate fire extinguishers mounted within 75 feet of outside areas containing flammable liquids, and within 10 feet of any inside storage area for such materials?

☐ Is the transfer/withdrawal of flammable or combustible liquids performed by trained personnel?

☐ Are fire extinguishers mounted so that employees do not have to travel more than 75 feet for a Class A fire or 50 feet for a Class B fire?

☐ Are employees trained in the use of fire extinguishers?

☐ Are all extinguishers serviced, maintained, and tagged at intervals not to exceed one year? Is a record maintained of required monthly checks of extinguishers?

☐ Are all extinguishers fully charged and in their designated places? Are extinguishers free from obstruction or blockage?

☐ Where sprinkler systems are permanently installed, are the nozzle heads directed or arranged so that water will not be sprayed into operating electrical switchboards and equipment?

☐ Are "NO SMOKING" signs posted where appropriate in areas where flammable or combustible materials are leased or stored?

☐ Are "NO SMOKING" signs posted on liquefied petroleum gas tanks?

☐ Are "NO SMOKING" rules enforced in areas involving storage and use of flammable materials?

□ Are safety cans used (for dispensing flammable or combustible liquids) at the point of use?

□ Are all spills of flammable or combustible liquids cleaned up promptly?

Hazardous Chemical Exposures

□ Is employee exposure to chemicals kept within acceptable levels?

□ Are eyewash fountains and safety showers provided in area where caustic corrosive chemicals are handled?

□ Are all employees required to use personal protective clothing and equipment (gloves, eye protection, respirators) when handling chemicals?

□ Are flammable or toxic chemicals kept in closed containers when not in use?

□ Where corrosive liquids are frequently handled in open containers or drawn from storage vessels or pipelines, are adequate means provided to neutralize or dispose of spills or overflows (properly and safely)?

□ Have standard operating procedures been established, and are they being followed, when cleaning up chemical spills?

□ When needed for emergency use, are respirators stored in a convenient, clean and sanitary location?

□ Are emergency-use respirators adequate for the various conditions under which they may be used?

□ Are employees prohibited from eating in areas where hazardous chemicals are present?

□ Is personal protective equipment provided, used and maintained whenever necessary?

□ Are there written *Standard Operating Procedures* for selecting and using respirators where needed?

□ If you have a respirator protection program, are your employees instructed on the correct usage and limitations of the respirators?

□ Are the respirators NIOSH-approved for each particular application?

□ Are respirators inspected and cleaned, sanitized, and maintained regularly?

□ Are you familiar with the Threshold Limit Value (TLV) or Premissable Exposure Limit (PEL) of airborne contaminants and physical agents used in your workplace?

□ Have you considered having an industrial hygienist or environmental health specialist evaluate your work operations?

□ If internal combustion engines are used, is carbon monoxide kept within acceptable levels?

□ Is vacuuming used rather than blowing or sweeping dusts whenever possible for cleanups?

Hazard Communication

□ Have you compiled a list of hazardous substances that are used in your workplace?

□ Is there a written hazard communication program dealing with Material Safety Data Sheets (MSDS), labeling, and employee training?

□ Who is responsible for MSDSs, container labeling, employee training?

□ Is each container for a hazardous substance (vats, bottles, storage tanks) labeled with product identity and a hazard warning (communicating with specific health hazard and physical hazards)?

□ Is there an MSDS readily available for each hazardous substance used?

□ How will you inform other employees whose employees share the same work area where hazardous substances are used?

Do you have an employee training program for hazardous substances? Does this program include:

☐ An explanation of what an MSDS is, and how to use and obtain one? An explanation of "Right to Know?"

☐ The contents of the MSDS for each hazardous substance or class of substances?

☐ Informing employees where they can review the employer's written hazard communication program, and where hazardous substances are located in work areas?

☐ Explaining the physical and health hazards of substances in the work area, and how to detect their presence, and specific protective measures to be used?

☐ Hazard communication program details including labeling system and MSDS use?

☐ How employees will be informed of hazards of nonroutine tasks, and hazards of unlabeled pipes?

Electrical Safety

☐ Are your workplace electricians familiar with the OSHA Electrical Safety Code?

☐ Do you require compliance with OSHA rules on all contract electrical work?

☐ Are all employees required to report (as soon as practicable) any obvious hazard to life or property observed in connection with electrical equipment or lines?

☐ Are employees instructed to make preliminary inspections and/or appropriate tests to determine what conditions exist before starting work on electrical equipment or lines?

☐ When electrical equipment or lines are to be serviced, maintained, or adjusted, are necessary switches opened, locked out, and tagged?

☐ Are portable hand-held electrical tools and equipment grounded or else are they of the double insulated type?

☐ Are electrical appliances such as vacuum cleaners, polishers, and vending machines grounded?

☐ Do extension cords have a grounding conductor? Are multiple plug adaptors prohibited?

☐ Are ground-fault circuit interrupters installed on each temporary 15 or 20 ampere, 120-volt AC circuit at locations where construction, demolition, modifications, alterations, or excavations are being performed?

☐ Are all temporary circuits protected by suitable disconnecting switches or plug connectors at the junction with permanent wiring?

☐ Is exposed wiring and cords with frayed or deteriorated insulation repaired or replaced promptly?

☐ Are flexible cords and cables free of splices or taps?

☐ Are clamps or other securing means provided on flexible cords or cables at plugs, receptacles, tools, equipment, and is the cord jacket securely held in place?

☐ Are all cords, cable, and raceway connections intact and secure?

☐ In wet or damp locations, are electrical tools and equipment appropriate for use or locations (or otherwise protected)?

☐ Is the location of electrical power lines and cables (overhead, underground, underfloor, other side of walls) determined before digging, drilling, or similar work started?

☐ Is the use of metal measuring tapes, ropes, hand lines, or similar devices with metallic thread woven into the fabric, prohibited where these could come into contact with energized parts of equipment, fixtures, or circuit conductors?

☐ Are all disconnecting means always opened before fuses are replaced?

☐ Do all interior wiring systems include provisions for grounding metal parts or electrical raceways, equipment, and enclosures?

□ Are all electrical raceways and enclosures securely fastened in place?

□ Are all energized parts of electrical circuits and equipment guarded against accidental contact by approved cabinets or enclosures?

□ Is sufficient access and working space provided and maintained around all electrical equipment to permit ready and safe operations and maintenance?

□ Are all unused openings (including conduit knockouts) of electrical enclosures and fittings closed with appropriate covers, plugs, or plates?

□ Are electrical enclosures such as switches, receptacles, and junction boxes provided with tight fitting covers or plates?

□ Are employees prohibited from working alone on energized lines or equipment over 600 volts?

□ Are employees forbidden from working closer than 10 feet of high-voltage (over 750 volts) lines?

Noise

□ Are there areas i your workplace where continuous noise levels exceed 85 dBA? (To determine maximum allowable levels for intermittent or impact noise, see OSHA's *Noise and Hearing Conservation* Rules.)

□ Are noise levels being measured using a sound level meter or an octave band analyzer and records of these levels being kept?

□ Have you tried isolating noise machinery from the rest of the operation? Have engineering controls been used to reduce excessive noise?

□ Where engineering controls are not feasible, are administrative controls (worker rotation) being used to minimize individual employee exposure to noise?

□ Is there an ongoing preventative health program to educate employees in safe levels of noise and exposure, effects of noise on their health, and use of personal protection?

□ Are employees who are exposed to continuous noise above 85 dBA retrained annually?

□ Have work areas (where noise levels make voice communication difficult) been identified and posted?

□ Is approved hearing protection equipment (noise attenuating devices) used by every employee working in areas where noise levels exceed 90 dBA?

□ Are employees properly fitted, and instructed in the proper use and care of hearing protection?

□ Are employees exposed to continuous noise above 85 dBA given periodic audiometric testing to ensure that you have an effective hearing protection system?

Identification of Piping Systems

□ When nonpotable water is piped through a facility, are outlets or taps posted to alert employees that it is unsafe and not to be used for drinking, washing, or personal use?

□ When hazardous substances are transported through above-ground piping, is each pipeline identified?

□ Have asbestos-covered pipelines been identified?

□ When pipelines are identified by colored paint, are all visible parts of the line well identified?

□ When pipelines are identified by color-painted bands or tapes, are these located at reasonable intervals, and at each outlet, valve, or connection?

□ When pipelines are identified by color, is the color code posted at all locations where confusion could introduce hazards to employees?

□ When the contents of pipelines are identified by name or abbreviations, is the information readily visible on the pipe near each valve or outlet?

□ When pipelines carrying hazardous substances are identified by tags, are the tags constructed of durable material, the message clearly and permanently distinguishable, and tags installed at each valve or outlet?

☐ When pipelines are heated by electricity, steam, or other external source, are suitable warning signs or tags placed at unions, valves, or other service-able parts of the system?

Materials Handling

☐ Are materials stored in a manner to prevent sprain or strain injuries to employees when retrieving the materials?

☐ Is there safe clearance for equipment through aisles and doorways?

☐ Are aisleways permanently marked, and kept clear to allow safe passage?

☐ Are motorized vehicles and mechanized equipment inspected daily or prior to use?

☐ Are vehicles shut off and brakes set prior to loading and unloading?

☐ Are containers of combustibles or flammables, when stacked while being moved, always separated by dunnage sufficient to provide stability?

☐ Are dock boards (bridge plates) used when loading and unloading operations are taking place between vehicles and docks?

☐ Are trucks and trailers secured from movement during loading and unloading?

☐ Are dock plates and loading ramps constructed and maintained with sufficient strength to support imposed loading?

☐ Are hand trucks maintained in safe operating condition?

☐ Are chutes equipped with side boards of sufficient height to prevent materials from falling off?

☐ Are chutes and gravity roller sections firmly placed or secured to prevent displacement?

☐ At the delivery end of rollers or chutes, are provisions made to brake the movement of materials?

☐ Are materials handled at a uniform level to prevent lifting or twisting injuries?

☐ Are material-handling aids used to lift or transfer heavy or awkward objects?

☐ Are pallets usually inspected before loading and/or moving?

☐ Are hooks with safety latches or other devices used when hoisting materials so that slings or load attachments won't accidentally slip off the hoist hooks?

☐ Are securing chains, ropes, chokers or slings adequate for the job being performed?

☐ When hoisting materials or equipment, are provisions made to ensure that no one will be passing under suspended loads?

Transporting Employees & Materials

☐ Do employees operating vehicles on public thoroughfares have operator licenses?

☐ Are motor vehicle drivers trained in defensive driving, and proper use of the vehicle?

☐ Are seat belts provided and are employees required to use them?

☐ Does each van, bus, or truck routinely used to transport employees have an adequate number of seats?

☐ When employees are transported by truck, are provisions provided to prevent their falling from the vehicle?

☐ When transporting employees, are vehicles equipped with lamps, brakes, horns, mirrors, windshields, and turn signals that are in good repair?

☐ Are transport vehicles provided with handrails, steps, stirrups, or similar devices that have been placed and arranged so employees can safely mount or dismount?

☐ Is a fully-charged fire extinguisher, in good condition, with at least "4 B:C" rating maintained in each employee transport vehicle?

☐ When cutting tools with sharp edges are carried in passenger compartments of employee transport vehicles, are they placed in closed boxes or containers which are secured in place?

☐ Are employees prohibited from riding on top of any load which can shift, topple, or otherwise become unstable?

☐ Are materials that could shift and enter the cab secured or barricaded?

Tire Inflation

☐ Where tires are mounted and/or inflated on drop center wheels, is a safe practice procedure posted and enforced?

☐ Where tires are mounted and/or inflated on wheels with split rims and/or retainer rings, is a safe practice procedure posted and enforced?

☐ Does each tire inflation hose have a clip-on chuck with at least 24 inches of hose between the chuck and an inline valve and gauge?

☐ Does the tire-inflation control valve automatically shut off the air flow when the valve is released?

☐ Is a tire-restraining device such as a cage rack used while inflating tires mounted on split rims or rims using retainer rings?

☐ Are employees strictly forbidden from taking a position directly over or in front of a tire while it is being inflated?

Emergency Action Plan

☐ Have you developed an emergency action plan?

☐ Have emergency escape procedures and routes been developed and communicated to all employees?

☐ Do employees who must remain to operate critical plant operations before evacuating know the proper procedures?

☐ Is the employee alarm system that provides warning for emergency action recognizable and perceptible above ambient conditions??

☐ Are alarm systems properly maintained and tested regularly?

☐ Is the emergency action plan reviewed and revised periodically?

Do employees know their responsibilities:

☐ For reporting emergencies?

☐ During an emergency?

☐ For performing rescue and medical duties?

Infection Control

☐ Are employees potentially exposed to infectious agents in body fluids?

☐ Have occasions of potential occupational exposure been identified and documented?

☐ Has a training and information program been provided for employees exposed to or potentially exposed to blood and/or body fluids?

☐ Have infection control procedures been instituted where appropriate, such as ventilation, universal precautions, workplace practices, and personal protective equipment?

☐ Are employees aware of specific workplace practices to follow when appropriate (handwashing, handling sharp instruments, handling of laundry, disposal of contaminated materials, reusable equipment, etc.)?

☐ Is personal protective equipment provided to employees, and in all appropriate locations?

☐ Is the necessary equipment (mouthpieces, resuscitation bags, other ventilation devices) provided for administering mouth-to-mouth resuscitation on potentially infected patients?

☐ Are facilities/equipment to comply with workplace practices available, such as handwashing sinks, biohazard tags and labels, sharp containers, and detergents/disinfectants to clean up spills?

☐ Are all equipment, and environmental and working surfaces cleaned and disinfected after contact with blood or potentially infectious materials?

☐ Is infectious waste placed in closable, leak-proof containers, bags, or puncture-resistant holders with proper labels?

☐ Has medical surveillance including HBV evaluation, antibody testing, and vaccination been made available to potentially exposed employees?
How often is training done and does it cover:
☐ Universal precautions?
☐ Personal protective equipment?
☐ Workplace practices which should include blood drawing, room cleaning, laundry handling, and cleanup of blood spills?
☐ Needlestick exposure/management?
☐ Hepatitis B vaccination?

Ergonomics

☐ Can the work be performed without eye strain or glare to the employees?

☐ Can the task be done without repetitive lifting of the arms above the shoulder level?

☐ Can the task be done without the worker having to hold his/her elbows out and away from the body?

☐ Can workers keep their hands/wrists in a neutral position when working?

☐ Are mechanical assists available to the worker performing materials-handling tasks?

☐ Can the task be done without having to stoop the neck and shoulders to view the work?

☐ Are pressure points on any part of the body (wrists, forearms, back of thighs) being avoided?

☐ Can the work be done using the larger muscles of the body?

☐ Are there sufficient rest breaks, in addition to the regular rest breaks, to relieve stress from repetitive motion tasks?

☐ Are tools, instruments and machinery shaped, positioned, and handled so that tasks can be performed comfortably?

☐ Are all pieces of furniture adjusted, positioned, and arranged to minimize strain on the body?

☐ Are unnecessary distances eliminated when moving materials?

☐ Are lifts confined within the knuckle to shoulder zone? Does the task require fixed work postures?

☐ Is work arranged so that workers are not required to lift and carry too much weight?

☐ If workers have to push or pull objects using great amounts of force, are mechanical aids provided?

Ventilation for Indoor Air Quality

☐ Does your HVAC system provide at least the quantity of outdoor air designed into the system at the time the building was constructed?

☐ Is the HVAC system inspected at least annually and maintained in a clean and efficient manner?

☐ Are efforts made to purchase furnishings or building treatments which do not give off toxic or offensive vapors?

☐ Are indoor air quality complaints investigated, and the results conveyed to workers?

Video Display Terminals (VDTs)

☐ Can the work be performed without eye strain or glare to the employees?

☐ Can workers keep their hands/wrists in a neutral position when working?

☐ Can the task be done without having to stoop the neck and shoulders to view the task?

☐ Are pressure points on any part of the body (wrists, forearms, back of thighs) being avoided?

☐ Are there sufficient rest breaks, in addition to the regular rest breaks, to relieve stress from repetitive motion tasks?

☐ Are all pieces of furniture adjusted, positioned, and arranged to minimize strain on the body?

☐ Are fixed work postures avoided in the task?

Recommended VDT Workstation Criteria

☐ Height of work surface: Adjustable 23 to 28 inches (58.4 to 71.1 cm).

☐ Width of work surface: 30 inches (76.0 cm).

☐ Viewing distance: 16 to 22 inches (40.6 to 55.8 cm) for close-range focusing.

☐ Thickness of work surface: 1 inch (2.5 cm).

☐ Eyes in relation to screen: Topmost line of display should be at approximately eye level (or lower for bifocal wearers).

☐ Blink rate: No more than two different blink rates, at least 2 Hertz (Hz) apart-slow blink rate not less than 0.8 Hz; fast blink rate not more than 5 Hz.

☐ Knee room width: 20 inches (51.0 cm) minimum.

☐ Knee room depth: Minimum of 15 inches (38.1 cm) knee level; 23.5 inches (59.7 cm) toe level.

☐ Seat height: Adjustable 16 to 20.5 inches (40.0 to 52.1 cm).

☐ Seat size: 13 to 17 inches (33.0 to 43.2 cm) depth; 17.7 inches (45.0 cm) to 20 inches (51.0 cm) width; "waterfall" front edge.

☐ Seat slope: Adjustable 0 degrees to 10 degrees backward slope.

☐ Backrest size: 15 to 20 inches high (38.1 to 50.8 cm); 13 inches wide (33.0 cm).

☐ Backrest height: Adjustable 3 to 6 inches (8.0 to 15.0 cm) above seat.

☐ Backrest tilt: Adjustable 15 degrees.

☐ Angle between backrest and seat 90 degrees to 105 degrees.

☐ Angle between seat and lower leg: 60 degrees to 100 degrees.

☐ Angle between upper arm and forearm in relation to keyboard: upper arm and forearm should form a right angle (90 degrees); hands should be in a reasonably straight line with the forearm.

Additional VDT Workstation Criteria

☐ Non-adjustable work surfaces: Table surface should be about 29 inches (73.6 cm) high with a key board surface height of 27 inches (68.5 cm).

☐ VDT stands: Height-adjustable stands for all new installations.

☐ Seats: Easily adjustable swivel chairs on five-point base.

☐ Footrests: If operator cannot keep both feet flat on floor when chair height is properly adjusted to work surface.

☐ Keyboards: Thin; detached from console; palm rest.

☐ Non-keyboard entry devices: Position devices following same guidelines for keyboards.

☐ Screens: Readable with no perceptible flicker; brightness control necessary.

☐ Blink rate: Nor more than two different blink rates, at least 2 Hertz (Hz) apart - slow blink rate not less than 0.8 Hz; fast blink rate not more than 5 Hz.

☐ Glare Control:

a) VDT screen placed at right angels to windows; screens have tilt and swivel adjustments.

b) Windows with curtains, drapes or blinds to reduce bright outside light.

c) Lighting levels at 30-50 footcandles when using a VDT; 50-70 footcandles where documents are read, compared to normal office levels of 75-160 footcandles.

d) Diffusers, cube louvres, or parabolic louvres to reduce overhead-lighting glare.

e) Work surfaces with anti-glare (matte) finish.

f) Movable task or desk lights; VDTs located between rows of overhead lighting; screen filters and/or hoods if above not successful.

☐ Cables and cords: Concealed, covered, or otherwise safely out of the way.

☐ Ventilation: Additional ventilation or air conditioning to overcome heat generated by more than one VDT workstation in the same room.

☐ Temperature and humidity: Maintain thermal comfort; 30-60 percent relative humidity.

☐ Noise: Acoustical enclosures for printers if sound levels exceed 55 dBA; main CPUs and diskdrives isolated.

☐ Training: Operators trained on how to adjust chair, workstation heights, screen brightness, and correct seat posture.

☐ Fatigue control: Good operator posture; body and eye exercises; rest pauses; job rotation or substitution of less demanding tasks.

☐ Vision problems: Evaluate operators who may need glasses or wear bifocals.

☐ Psycho-social issues: Operator involvement in selection process; communication between operators and supervisors; user-friendly software, and adequate operator training.

Cranes & Hoists

☐ Are cranes visually inspected for defective components prior to the start of any work shift?

☐ Are all electrically-operated cranes effectively grounded?

☐ Is a crane preventive maintenance program established?

☐ Is the load chart clearly visible to the operator?

☐ Are all operators trained, and provided with the operator's manual for the particular crane being operated?

☐ Have construction industry crane operators been issued a valid operator's card?

☐ Are operating controls clearly identified?

☐ Is a fire extinguisher provided at the operator's station?

☐ Is the rated capacity visibly marked on each crane?

☐ Is an audible warning device mounted on each crane?

☐ Is sufficient lighting provided for the operator to perform the work safely?

☐ Are cranes with booms that could fall over backward, equipped with boomstops?

☐ Does each crane have a certificate indicating that required testing and examinations have been performed?

☐ Are crane inspection and maintenance records maintained and available for inspection?